# Construction Contracting

**James J. Adrian, Ph.D., PE, CPA**
Professor, Bradley University
President, Adrian International L.L.C.

**Douglas J. Adrian, CPA**
Vice President, Adrian International L.L.C.

ISBN 0-87563-938-0

Published by

**Stipes Publishing L.L.C.**
204 W. University Ave.
Champaign, Illinois 61820

# Preface

The construction industry represents one of the largest and most challenging of all industries. Each and every day, new buildings, roadways, and houses are constructed in virtually every city. Construction is the backbone of the economy. The construction industry is a growth industry. Projects wear out and need to be rebuilt. In addition, new construction is also demanded by social, technological, economic, and demographic changes.

The construction contractor building projects has to be hard working, creative, ethical, and experienced. Think about the challenge of constructing a project. What would your car cost if it were built under the following conditions: It is a unique design. It is built on your driveway, subject to extremes of temperature as well as wind and rain and snow. It is assembled by more than ten different firms representing more than ten different trades — each one maneuvering for space just to work. Several hundred different types of materials are being delivered to your driveway. In addition, a small flotilla of government inspectors are milling around the jobsite, poking their fingers here and there, scrutinizing the entire project for manufacturing defects. How much would the car cost? Would it run?

Obviously I have just described the construction process. The construction contractor encounters these difficulties on every project. In addition, given the large number of construction contractors, the construction firm must construct a project in a very competitive marketplace. It is not unusual for the construction contractor to be bidding against five or more contractors for a single project.

In this book, we cover the business and project management functions that are necessary to being a successful construction contractor. After an introductory chapter, chapters two through eight cover the office or company business functions of a successful contractor. Chapters include marketing, financing, accounting, and business law.

Chapters nine through thirteen cover project management functions — functions necessary to constructing a successful project. Topics include estimating, planning and scheduling, productivity measurement and improvement, control, and change order and claims management.

The appendix of the book includes checklists for the contractor to evaluate and improve their project management functions. The use of these

i

checklists provides the firm a means of setting out practices to improve their project management operations.

The intended audience of this book is both the practitioner and the student. The practitioner should be able to use the book as both a company and project primer — a handbook for construction contracting. The book should provide the student with an understanding of the construction contracting environment and provide the student the skills to work for a construction contractor, or the skills to start his or her own firm.

We would like to thank Sandy for her review of the manuscript. We also are thankful to all our industry contacts that have indirectly educated us and provided us the information for this book.

Peoria, March 2000

Dedicated to Sandy

# Contents

# List of Figures

# Chapter 1

# Successful Construction Contracting: An Introduction

## 1.1   Construction Industry Environment

The contracting business is characterized by significant opportunity for improvement and significant potential profits. The construction industry itself is one of the largest of all U.S. industries. Annually, there is approximately 400 to 800 billion dollars of new construction funded in the U.S. Annually, new construction expenditures account for 7 to 12 percent of the Gross National Product (GNP) of the United States. The annual expenditures for new construction are divided almost equally between three segments of the industry:

- Residential construction
- Building construction
- Heavy and highway or public works construction

The construction industry, to include contracting, is an ongoing growth industry. Given the limited life of construction materials to include concrete, wood, and steel, projects have to renovated and rebuilt. Maintenance and repair of existing construction projects is another segment of annual construction expenditures.

While construction represents one of the largest of all U.S. industries, competition for work amongst construction contractors is extremely competitive. There are in excess of 500,000 firms that can be classified as construction firms. It follows that there is a limited amount of construction work to be done by each firm. For example, if one assumes that there is $500 billion dollars of new U.S. construction in any one year, and that there are

500,000 firms competing for this work, it follows that there is $1,000,000 of work to be done by each firm. Obviously some construction firms do more than this volume, others do less.

While there are several large international type construction firms that do large volumes of work, the majority of construction firms are small, closely held businesses. Many firms are one person owner firms that operate on low overhead and as such are very competitive in obtaining work.

Construction volume or activity is also characterized as being quite variable from one time period to the next. Construction expenditures are dependent on many factors to include the following:

- Governmental expenditures for new construction
- Monetary policy; as interest rates rise, construction expenditures drop
- Tax laws that favor or disfavor new privately funded projects
- The needs of individuals and business
- Natural disasters that create the need for new construction

All of the above factors that in part dictate construction expenditures vary as a function of time. The end result is that construction expenditures and the construction volume of the contractor vary considerably from one time period to the next.

Each and every construction project, to include the small and large, entails the controlling of four project variables:

1. Cost of the project
2. Duration of the project
3. Quality of the project
4. Safety of the project

The construction industry is part of the manufacturing industry. The contracting firm can be thought of as a firm that uses labor and equipment to place material into a project. The firm's ability to generate profit and be successful is dependent on the firm's ability to utilize labor and equipment efficiently. The importance of the utilization of labor and equipment is evident when one analyzes the cost components of the typical construction project.

As can be observed in Figure 1.1, the contractor is very dependent on labor cost and equipment cost. It follows that the successful contractor must use labor and equipment efficiently. A measure of the efficiency of using a resource is referred to as productivity.

The low profit margin shown (2%) is dictated in great part by the competitive nature of the construction industry. In a competitive industry, the industry at large dictates this low profit margin. The two major profit centers for any construction firm are labor and equipment. Productivity dictates these two cost components. Productivity is defined as follows:

$$\text{Productivity} = \frac{\text{Units of work placed}}{\text{Person hours of effort}}$$

|                                   | Building  | %   | Highway   | %   |
|-----------------------------------|-----------|-----|-----------|-----|
| Direct Labor Cost                 | 400,000   | 40  | 250,000   | 25  |
| Direct Material Cost              | 350,000   | 35  | 170,000   | 17  |
| Direct Equipment Cost             | 80,000    | 8   | 400,000   | 40  |
| Job Overhead — General Conditions | 80,000    | 8   | 80,000    | 8   |
| Company Overhead                  | 70,000    | 7   | 80,000    | 8   |
| Profit                            | 20,000    | 2   | 20,000    | 2   |
| Total                             | 1,000,000 | 100 | 1,000,000 | 100 |

**Figure 1.1.** Typical components of contractor bid.

While the definition has person hours in the denominator, one could argue that it should also include equipment hours. To be profitable, a contractor has to use equipment and labor efficiently to increase productivity.

The construction industry can be characterized as an industry that has been slow to increase productivity. Many studies indicate that as much as fifty percent of an eight-hour day at a job site can be considered non-productive time. An example analysis of an eight-hour work day for on-site labor is illustrated in Figure 1.2.

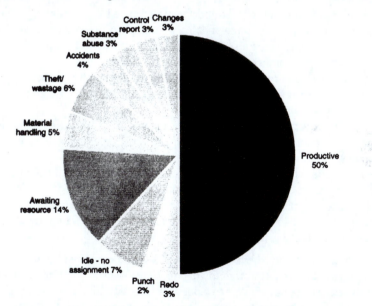

**Figure 1.2.** Example labor work states during 8-hour day.

The non-productive time shown can be considered as opportunity to improve potential profit centers for the construction firm. Through effective on site management, the contractor can eliminate non-productive time. Given the data shown in Figure 1.2, it follows that the elimination of unnecessary handling of material would increase productivity by five percent. A five percent improvement would reduce on-site labor costs by five percent. This reduction would have the effect of doubling profits.

An analysis of equipment at the job site would likely result in a determination that equipment non-productive time would be even greater than that shown for labor in Figure 1.2. It follows that the management of labor and equipment productivity is the key to a profitable construction firm.

## 1.2    The Construction Contractor

The contractor often performs specialized work. The firm has several unique characteristics, some of which are favorable and some of which are unfavorable. Some of these unique characteristics are the following:

- The work is very labor intense.
- The work entails placing concrete, the placement of steel, and many other types of work. Each of these tasks is interdependent; if one is performed inadequately, it affects the cost, duration, quality, or safety of the overall project.
- There are often many ways to perform a required work task; e.g., there are numerous work systems that can be used to do a task. This characteristic enables the contractor to be creative and gain a competitive edge by working smarter, not necessarily harder.
- The business is very competitive. The relatively small capital required to start the firm has resulted in many contracting firms all competing for the same work.
- There has been many new products produced by manufactures that enable the contracting company to improve the time, duration, quality, or safety of the project.
- Given the difficulty of the work process, accident rates and the severity of accidents in construction work are significant.
- The various types of work result in varying degrees of quality of workmanship. It follows that labor skills are critical to achieving a high quality contracting project.

While some of the above characteristics can be viewed as being unfavorable, the fact remains that owing to the potential for improvement, the contracting business is a great business to start and a great business in which to generate profits. Using hard work, creativity, business as well as technical skills, and honesty and ethics, an individual can attain great success in the contracting business. Each of the issues listed in Figure 1.3 can be viewed as having a high degree of potential improvement.

| Potential for Improvement | Comments |
| --- | --- |
| On-site labor productivity | Studies conducted by the author indicate that on-site labor is in a non-productive state almost fifty percent of the time owing to reasons such as redo work, waiting, double handling material, etc. A small increase in productivity can significantly increase the contracting firm's profits. |
| Equipment use and productivity | Many contracting firms get less than 900 productive hours of use from their equipment in a given year. Improved equipment management can aid cost as well as duration project variables. |
| Accidents | The construction industry, to include the contracting business, has one of the highest rates of accidents per labor hour worked. A negative impact of a single accident affects project time, cost, as well as the negative outcome to the worker. There is no excuse for having a single accident on the project. |
| Planning and scheduling | Studies indicate that many contracting firms, to include contracting companies, construct projects without having a formalized project schedule. Constructing a project without a formalized schedule is like driving to a new location in a car and not taking a roadmap. The trip will take more time, cost more money, and will be accompanied by disputes and unpleasantness. The contracting company can improve project time, cost, and quality through improved scheduling practices. |
| Estimating | Many construction firms, to include the contracting company prepare estimates from "the seat of their pants"; a less than defendable approach. Improved historical data bases regarding labor productivity, quantity take-off procedures, and overhead allocation procedures are just a few of the many ways the firm can improve the risk taking characteristic of the work process. |
| Accounting and financial procedures | Inadequate accounting and financial management practices, to include inadequate or untimely job site record keeping, can prevent the contracting company from identifying fixable on-site project problems. While there are excuses for inadequate record keeping (e.g., it may be difficult to get production-type people (foremen) to be attentive to the importance of the accounting function), inadequate accounting and financial management lead to non-optimal project cost, time, quality, and safety. |
| High insurance and legal costs | Historically, the construction firm has had to expend considerable funds on obtaining insurance for both a project and the overall firm. In addition, the firm may incur considerable legal expenses related to project disputes. Neither insurance nor legal costs provide added value to a project or to the firm. The reduction of these costs would benefit the project variables of cost, time, quality, and safety. |

**Figure 1.3.** Potential for improvement.

The issues in the left hand column in the table above can be viewed either as problems or opportunities. If one recognizes that each and every contracting firm has the problems listed above, the individual firm should take the view that each of these is an opportunity. Improvement of any of the above issues gives the firm an edge over its competition; an edge that can relate to a potential for significant profits. The opportunities for improvement are significant. This is the most favorable attribute of entering the contracting business.

## 1.3    The Construction Contractor as a Small Business

Each year, thousands of individuals, motivated by initiative and high expectations, launch a firm in search of fulfilling their dream of success. Tired of the corporate politics, committees, and a feeling of being lost in the shuffle, they pursue their dream of "rags to riches". These thousands of individual firms have a major factor in the American business system and the construction industry.

The Small Business Administration (SBA) identifies the construction firm as a small business if the firm's average annual volume of work is less than five million dollars for the preceding three years. The majority of contracting firms fit this definition.

There are in excess of 9 million small business entities in the United States. This represents more than 95 percent of all business entities. Small businesses employ approximately fifty percent of all employees in the U.S.

Small businesses, to include a contracting firm, offer the owner significant rewards. The primary incentive for the small business entrepreneur is that, being his own boss, he is working for himself. Thus the profit derived from the firm operations serves as the financial reward. To a large extent, the owner of the firm controls the amount of his income by the type of decisions he makes, by the effort he makes, and by his technical and managerial expertise. Other advantages include the close contact the owner has with his employees and customers, the ease of entry (and departure) into business, and the pride of ownership.

There are also disadvantages of owning a small business. The owner will typically work long hours. Very often he is the first to arrive in the morning and the last to leave at night. If he becomes ill, he cannot call in sick to his boss; he is the boss. He seldom can take a few days off unless he has dependable and trustworthy employees.

As a small firm, he must also "do it all"; he must be a jack of all trades. In a large business, specialists are hired to perform critical functions. Accountants maintain the financial records. Marketing specialists solicit and maintain clients. Production and technical specialists, such as engineers, oversee the manufacturing and design functions. Thus a major drawback to many small firms is their difficulty in acquiring and utilizing specialists.

The owner of the small firm may also feel isolated. Unlike working for a large corporation, where individuals can bounce problems off colleagues and committees, the owner of the small firm takes all problems on his own shoulders. He must bear the good with the bad.

Perhaps the biggest disadvantage to a small business is the threat of financial failure. The number of business failures in the United States peaked in 1932 at over 30,000. This was during the country's economic depression. The lowest number, less than 2,000, occurred in 1945. The number gradually increased to over 17,000 in 1961 and has been on a slight decline every since. While business failures always threaten a business, one should put this into perspective. The number of business failures is currently around 30 to 40 for every 10,000 businesses. This represents only a small percentage. Retail businesses have the highest rate of failure, in excess of 60 for every 10,000 started. If one omits this industry from the totals, the failure rate can be considered small.

Nonetheless, business failure, is a potential threat to the small business, to include the contracting firm. By analyzing the causes of business failures, the contractor can adequately prepare himself in the management and technical knowledge and skills necessary to prevent financial problems.

Each year, Dun & Bradstreet, makes available information relating to causes of business failures. This data for a recent year is illustrated in Figure 1.4.

| Underlying Causes | Percent of Total Failures |
| --- | --- |
| Neglect | 1.6 |
| Fraud | 1.3 |
| Lack of technical experience | 16.4 |
| Lack of managerial and business experience | 14.1 |
| Unbalanced experience | 21.6 |
| Incompetence | 41.0 |
| Disaster | 0.6 |
| Other | 3.4 |

**Figure 1.4.** Causes of business failures.

The majority of business failures were directly related to the owner's lack of technical knowledge, managerial and business knowledge and experience, and incompetence. Only a small percentage of the failures are attributed to neglect, fraud, or disaster.

Construction contracting, to include the small and large firm, has it's own unique characteristics that lead to specific financial difficulties. The author, in serving and observing numerous construction firms over a thirty year period, has accumulated his own list of why construction firms fail. The list in terms of frequency is shown in Figure 1.5

| Frequency | Cause of Financial Failure — Construction Firms |
|---|---|
| 1 | Inadequate project controls |
| 2 | Unexpected low productivity on a project |
| 3 | Inaccurate estimate |
| 4 | Lack of cash flow — liquidity |
| 5 | Lack of technical skill |
| 6 | Inability to collect receivables or construction claim |
| 7 | Lack of workload or revenue from sales |
| 8 | Inadequate project planning and scheduling |
| 9 | Lack of business controls on firm operations |
| 10 | Fraud, disaster, no insurance on fire, accident not covered |

**Figure 1.5.** Causes of contracting failures.

The number of business failures, or the list of reasons for business failures should not discourage the contractor. The firm should learn from the lists shown in Figures 1.4 and 1.5. By taking actions and gaining knowledge that offset these reasons for failure, the firm can reap the benefits of owning a small, or for that matter, a large contracting company. Risk takers are frequently rewarded with significant benefits; financial, and psychological.

## 1.4   Technical, Business, and Managerial Skills for the Contractor

As noted in the previous section, the contractor must have diverse skills, both technical and business skills. The technical knowledge of materials to include the properties of concrete, knowledge of methods of construction, and knowledge of alternative tools and equipment for performing work are critical to operating a successful contracting company.

The technical skills and the technological aspects of construction contracting are covered in detail in chapters in this book. The technical skills and experience of the contractor must be accompanied by business management skills. The competitive nature of the business make it crucial that the firm have a competitive business edge.

Business skills for the contracting firm can be divided into two somewhat separate categories; skills necessary for operating the firm, and skills required to construct projects. One of the unique features of the construction firm is that the profit center is their projects. Profits are generated through the performance of a series of individual projects. Firm operations can be viewed as supporting the construction of projects.

The business skills relevant to operating the contracting firm are common to all private firms. These skills are summarized in Figure 1.6.

| Business skills for managing the firm | Comment | Coverage in this book |
|---|---|---|
| Organizational structure of the firm | This includes determining the organizational structure of various individuals and departments, the legal (e.g., proprietorship versus corporation), and financial structure. | These topics are covered in Chapter 2. |
| Marketing and sales | This includes efforts made to locate, evaluate, and secure projects. | Chapter 3. |
| Finance | The contracting company, needs to secure short-, intermediate-, and long-term funds to carry out their business. Several sources of these types of funds are available to the firm. | Financing is covered in Chapter 4. |
| Financial accounting | Critical to any firm is the performance of daily accounting aimed at measuring the financial success of the firm. The financial accounting function has the main objective of producing a balance sheet, income statement, and supporting schedules. These financial statements are used by the firm itself and many external entities that evaluate the firm's performance; e.g., banks and surety companies. | The handling of daily financial entries is presented in Chapter 5. The preparation of financial statements is covered in Chapter 6. |
| Financial management | The ability to monitor and evaluate the financial structure and operation of the firm is performed by reviewing various financial ratios and indicators. This analysis is critical to detecting strengths and weaknesses of the firm. | Financial management is covered in Chapter 7. |
| Legal | Contract law includes an understanding of the legal requirements for a contract, the obligations of the parties, and how a contract is terminated. The contractor is also involved in negotiated paper transactions and in employment law. | Legal aspects of contracting are covered in Chapter 8. |

**Figure 1.6.** Business skills relevant to operating the contracting firm.

The topics covered in Figure 1.6 can be viewed as "firm" oriented business topics. In addition to these business skills, the contractor should understand and implement sound business practices related to constructing projects. While many skills are important, those listed in Figure 1.7 are essential to a successful contracting firm.

| Business skill for constructing a project | Comments | Coverage in this book |
|---|---|---|
| Estimating and bidding | The contractor expends much time and effort in preparing bids for project. This entails performing many estimating steps efficiently and accurately. These steps include quantity take-off, costing required labor hours, material, equipment, determining overhead amounts, and developing a profit strategy. | Estimating is covered in Chapter 9. |
| Planning and scheduling | The estimate for a project can be viewed as forecast of cost. It is important that the contractor have a plan for time. The planning and scheduling function includes determining the needed resources required to construct a project as a function of time. | Planning and scheduling are covered in Chapter 10. |
| Productivity measurement and improvement | The two most significant profit centers in any construction firm are labor and equipment productivity. There is a significant potential to improve both in the contracting business. Productivity is defined as the amount of work produced per person hour of effort. Critical to the improvement of productivity is the measurement of productivity. | Productivity measurement and improvement are covered in Chapter 11. |
| Project control | The estimating and planning and scheduling functions set out the plan for project time and cost. It is the control function that enables these to be realized. Control is the timely comparison of actual events to planned events with the potential to detect a problem and the follow up action that is taken to fix the problem. For the contracting firm this entails job site record keeping and the preparation of effective time job cost reports. | The project control function is covered in Chapter 12. |
| Construction claims and disputes | The nature of the construction process is that things don't always go as planned. Uncertain events may lead to time and cost disputes with the project owner or designer. The contractor's ability to prevent disputes and prepare information necessary to support a dispute can be crucial to the firm's profitability and going concern. | Disputes and claims are covered in Chapter 13. |

**Figure 1.7.** Skills essential to a successful contracting firm.

Clearly the successful contractor must possess a diverse set of technical as well as business skills. While the owner of the firm cannot be expected to be an expert in each and every one of the business skills set out above, a basic if not thorough understanding of each is necessary to operating a successful contracting firm.

# Chapter 2

# Structure of the Firm: Legal, Organizational, Financial

## 2.1 Introduction

A construction firm is one of the easier businesses in the United States to launch. The use of borrowed funds, the ease of licensing procedures, and the opportunity to rent needed equipment all stimulate entrance into the industry. Approximately 7% of all firms in the industry are new entrants annually.

Whereas the starting of a contractor firm is relatively easy, the starting of a firm that will note eventually fail is difficult. To accomplish this, the individual or individuals starting a contractor firm must be aware of all the facts which will effect the business at the start and in the future. Among these factors are the structure of the firm as to the type of work the firm performs, and its legal, organizational, and financial structure. The relevance of the alternatives available to the firm as to its structure are equally important throughout the life of the firm. Whereas one form of structure may be appropriate for the starting of the firm, another structure may be appropriate as the firm grows in size and volume of work performed. The alternative types of structure of the contracting firm and the advantages and disadvantages of each are discussed in this chapter.

## 2.2 Types of Construction Firms

The size of the firm and the annual volume of work it undertakes are constrained by the resources available to the firm and its bonding capacity.

Perhaps the best measure as to the size of a firm is its bonding capacity. Since the firm is required to submit a bond to the owner before undertaking a project, its bonding capacity limits the size of projects it can undertake and thus plays a role in the annual volume of work the firm performs.

More often than not, the contracting firm starts as a small firm. One classification system used classifies a small firm as a firm doing less than one million dollars of work volume annually. The break point between a middle or average sized firm and a large firm can be thought of as an annual work volume of twenty million dollars. Whereas some of the existing construction firms started as small firms and have increased their bonding capacity enough to be classified as medium or large firms, many other firms have continued to operated as small firms. Not every firm can or should substantially increase its size or annual volume of work. Many a firm has gone bankrupt trying to carry out its objective of joining the "top 400" firms.

The larger firm does have certain advantages in regard to the construction industry. Foremost among these advantages is the fact that on a percentage basis, the overhead per dollar value of work performed is less than that for the smaller firm. This is evidenced by comparison of income statements for the small and large firms. This ability to operate with a lower percentage overhead results in the larger firm being more competitive and being able to obtain a somewhat higher profit margin. Since the larger contractor can always take on projects that are less than its bonding capacity, it often is in a position to pick and choose from available projects rather than to have to seek to be low bidder on each and every project. The size of the firm relative to its capital structure may also result in the large firm being somewhat less sensitive to economic slowdowns than is the smaller firm.

The small firm is not without its advantages. Normally, it does not have the management problems that are characteristic of a firm that is large and increasing in size. In addition, the rewards of operating successfully are received by the owner and not shared with stockholders or investors as is typical in the case of the large firm. Naturally, the reverse is also true in regard to losses the firm might absorb. However the small firm, because of its relatively low dollar operating overhead, may be in a position to "shut down the shop" during a slowdown.

In general, the size of the firm is not dictated alone by its technical skills. Its growth has to be accompanied by growth in management skills and changes in its financial makeup. While not always the case, a bonding company will consider all three factors of technical skills, management ability, and financial soundness in increasing or decreasing a firm's bonding capacity.

# 2.3   Legal Structure of the Firm

The legal structure of the firm affects the everyday operations of the construction firm. The time spent seeking out the best legal form for the business typically pays for itself. The following broad forms of legal structure are available to the construction firm.

1. Proprietorship
2. Partnership
3. Corporation

Additionally, three modifications of the above types are relevant to the structure of the construction firm. These are as follows.

1. Limited Partnership
2. Joint Venture
3. Subchapter S Corporation

Each of these six legal structures, with a few exceptions, are available to the contracting firm when it is started and throughout the period of time it carries out its operations.

The decision as to the legal form chosen should recognize the following factors.

1. The effect on the management decision making process.
2. The effect on the income tax expense.
3. The liabilities of the owners of the firm.
4. Provisions for continuance of the firm.
5. The effect on the firm's ability to obtain funds
6. The initial cost to form the legal structure.
7. The degree of governmental control on the firm.

Other factors are relevant to a lesser degree in that they relate to specific firms. These seven factors as they relate to possible legal structure of the construction firm will now be discussed under the three broad classifications of legal structure.

## Proprietorship

The majority of very small construction firms are proprietorships. The arguments for such a legal structure center around retaining ownership and management of the firm for the owner of the firm, and the ease and low cost of taking on such a legal structure. To a lesser degree, the fact that there is little governmental control as to a proprietorship is another plus. On the negative side, the proprietorship is normally unfavorable in regard to tax expense, liability of the owner, ability to raise funds, and continuance of the firm in the case of death of the owner.

Typically, the owner of a contracting firm is somewhat suspicious about "opening up" his books to anyone but himself. This includes keeping his cost data and financial data confidential. In addition, he tends to believe in his own management ability and thus avoids outside help in the form of consultants. Much of this unwillingness to confide in others has no doubt developed from the fierce competition that is characteristic of the construction industry and the competitive nature of the bidding process in itself. The end result is that the contracting firm owner often rejects partnership or corporation legal structure on the basis that it defeats his confidentiality of business records and management know-how. It should be pointed out that this confidentiality can work unfavorably as to the contractor. In many cases, his business records and management practices are less than adequate. His confidentiality and ignorance of his failures often go unexposed when the firm is a proprietorship.

More often than not, the contracting firm is started under the proprietorship legal structure. This is due in part to the owner's desire to retain the management of the firm. However, an equally or more important factor in the popularity of the proprietorship form of legal structure is its ease of formation. Of all the alternative legal structures, the proprietorship is the easiest to form as to legal requirements or initial cost. Basically, anyone can start a proprietorship. Without the aid of legal counsel, an individual can buy himself a pick-up truck and call himself a general contractor or subcontractor. As to the cost, it is minimal. Whereas it may prove beneficial to consult legal or financial counsel, it is not necessary and this cost can be avoided. Some states require that a firm be licensed in order to be able to contract for construction work in the state. Whereas some states use this license fee as a revenue raising tool, others in fact use it as a means of controlling entrance of non-reputable firms. The point is that if in fact a state requires a license, it is required of the firm regardless of their legal structure. Even when such a license is required, the cost is minimal in that it typically is less than one hundred dollars. If such a license is required, it serves as the only external control on the operations of the proprietorship.

The disadvantages of the proprietorship legal structure grow as the firm grows. However, these disadvantages may in fact outweigh the advantages even when the firm first opens its doors. Perhaps foremost as to these disadvantages are the unlimited liability of the proprietor, and the unfavorable tax position of the proprietor. The liability of the proprietor is such that he risks both his business and personal assets when entering into a business liability. If debtors seek repayment of business loans, should the proprietor's business assets be insufficient to cover these loans, the debtor has a legal right to the proprietor's personal assets. In reality, little difference is attached to the business or personal assets of the proprietor.

The proprietor's income from his business becomes part of his personal income for determination of income tax expense. The disadvantage of this combining of business and personal income for the proprietor is discussed by means of examples in following discussions of alternative legal structures. Finally as pointed out earlier, typically the proprietorship is terminated upon the death of the proprietor. Through legal counsel, steps can be taken to provide for a continuance of the business upon the death of the proprietor (e.g., the firm transferred from a father to his son). However, the avoidance of such legal costs is viewed as one of the reasons for forming a proprietorship.

A partnership is a business with co-owners organized to make a profit. Approximately 10% of all construction firms are partnerships. The partnership legal structure is merely an extension of a proprietorship to two or more owners. However, because of multiple ownership, several characteristics arise that are unique in regard to each partner and to the external parties dealing with the partnership.

The incentive behind the partnership is "sharing" and "team effort". The belief of the partnership is that the sum of two parts will be more effective than each part taken individually. As such, individuals or firms may form a partnership with the objective of expanding and improving their areas of work and services offered, their management skills, or their financial structure.

Similar to the proprietorship, the partnership can be formed without detailed procedures or cost. While the partnership agreement does not have to be written, it is often advantageous to do so in that it helps in resolving potential disputes. While not severe, external constraints on the firm are somewhat greater than that of the proprietorship. For example, a contracting firm may have to disclose the names of the partners when carrying out its operations.

There are several unique operational characteristics of a partnership. These characteristics often lead to disputes among partners and external parties. In particular, confusion and disputes may result as to partnership income distribution, income tax effects on partners, individual partner management rights, liabilities of individual partners, and the effect of dissolution and windup (i.e., termination) of the partnership as to individual partners. Typically, these disputes are resolved by reference to the Uniform Partnership Act (U.P.A.).

Unless otherwise agreed to in the partnership agreement, partnership income is distributed equally to the partners. This is true even if one partner has contributed more capital than the other partner. An agreement to distribute income in an unequal ratio overrides the equal distribution rule.

It is important to note that the partnership legal structure serves merely as a funnel of income to individual partners. The partnership income is not taxed to the partnership. Rather all income is funneled to partners who are then taxed as individual taxpayers. For this reason there is an internal revenue provision that requires that the income distributions to partners be made in a manner that does not attempt to manipulate tax payments. For example, it is not permissible to distribute all capital gain income (which are taxed at a lesser rate) to a high income partner, and all ordinary income to a relatively low income partner for the purpose of minimizing total tax expense. Although the internal revenue permits good freedom in regard to partnership income distribution, they are particularly in disfavor of distribution plans that are not consistent from one year to the next. Because of the almost infinite combinations of possible ways of distributing partnership income, it is absolutely necessary that the agreement be clear and be such that it covers all possible amounts to be distributed.

As previously indicated, partnership income is distributed to the partners who in turn include their distributions in their individual tax returns. Thus the partnership income may be taxed at different rates.

When a contracting firm enters a partnership it is often not without apprehension as to the future management of the firm. Questions as to who has authority to do what often lead to disputes among the partners and eventual termination of the partnership. Much of this apprehension and dispute could be avoided if the management responsibility and authority of each partner is understood by all when the partnership is formed.

Unless the partners agree to the contrary, each partner has equal rights in the management of the operations of the partnership. When a management decision is made, the majority rules. However, a few decisions need the consent of all of the existing partners.

A partner is essentially a fiduciary. The partner has a duty to the partnership to act with loyalty and to account for his duties. For example, the partner has a duty not to devote excessive time to another business.

A partner has express, implied, and apparent authority. Expressed authority is seldom disputed in that it is set out in the partnership agreement. For example, a partner to a construction firm may have expressed authority to hire and fire personnel. Implied authority is the authority due to his being a partner in a certain type of business. As to a construction firm, implied authority means that a partner would have authority to sign contracts to build projects and purchase material. Apparent authority is the authority that an external party would reasonably assume a person would have because he was a partner in a given type of business. For example, even though a partner of a construction firm may not have express or implied authority to purchase a major piece of equipment, an equipment dealer may reasonably assume the partner has authority and as such the equipment dealer would have legal remedies should the purchase be disputed.

Given the potential for disputes and differences of opinions between partners, the disadvantages of such a legal structure for the contractor often outweigh the advantages. On the other hand, occupations and businesses such as lawyers, architects, and practicing engineers often choose the partnership legal strict in that one of the alternative forms, the corporation structure may not be available to them. Clients of these professions require a high degree of trust in that the owners of the business deal in personal services. Under the corporate structure, the actual owners may be unknown to the client and the element of personal trust is absent. On the other hand, the ownership, responsibility, and liability of the partnership are clearly recognized by clients of the partnership. As such, lawyers, architects, and practicing engineers may be forced into forming partnerships in order to gain the trust of clients.

Whereas there are few construction firms that choose the true partnership legal structure, there are several that may take a legal form that approaches such a legal structure. In particular, the limited partnership and the joint venture are legal structures similar to the partnership that are commonly used by construction related firms.

Many contracting firms have an unfavorable cash position as to short-term financing and cash for growth. The end result is that the firm may be willing to take on a limited partner. The limited partner invests cash or non-monetary assets in the firm for the right to a percentage of future profits. However, the limited partner has no management input to the firm. This retention of management decisions is compatible with the desires of the contractor who is in effect the general partner of the newly formed partnership.

From the limited partner's point of view, the arrangement is favorable in that he is offered a share of potential profits with only limited liability. Unlike a general partner, the limited partner's liability is limited to his investment.

As is true of any partnership, all parties involved in the general partner/limited partner arrangement should be fully aware of the implications of the agreement. For example, the limited partner has no authority as to signing contractors, making debts, etc., under the limited partnership agreement. As to third parties, it is unlawful to disclose the names of limited partners in the partnership name. Such a disclosure would lead external parties to believe that limited partners are in fact owners and thus fully liable for the debts that may come due to external parties.

The joint venture is not legally a partnership. It is an arrangement whereby two firms agree to "join forces" to carry out operations for a relatively short period of time or for a single project. The arrangement is a familiar one in the construction industry. A contracting firm may not have the technical capabilities or the bonding capacity necessary to bid successfully on a construction project. Thus, they seek out a compatible firm willing to team up for the project. The arrangement often proves advantageous to both parties (i.e., more than two parties can agree to a joint venture) in that individually neither may be in position to bid or make a profit on a project.

The joint venture is not without its potential difficulties. Foremost is the potential dispute as to who is responsible for completing various work items. In order for the joint venture to be successful, work responsibilities must be clearly defined. In addition, the arrangement must be truly a team effort. That is, each firm must be willing to give and take as to responsibilities, liabilities, and profits.

To the extent that each firm of the joint venture has control over its activities, it is liable for the debts of the joint venture. Individual firms to the joint venture have very limited powers to affect the contractual relations of each other. While typically there is a single contract with the project owner, variations are numerous in regard to the joint venture arrangement. In order to reduce or eliminate possible disputes that occur because of the variations, the construction firm typically enters into a joint venture with a firm that they have had a similar arrangement with on previous projects.

## Corporation

A contracting firm's decision to incorporate normally centers around potential income tax savings and a more favorable liability position for the owners of the company. Of less significance, but also favorable in the decision to incorporate are the considerations of obtaining capital funds, continuance of the firm, and allowable fringe benefits such as pension plans for officers.

The tax consideration is not favorable for every firm. If a firm is small and thus has little taxable income or the firm wishes to limit its income to single taxation, a corporate structure should not be chosen. In addition, a corporate legal structure implies a decentralizing of company management. While it is possible for a single owner to retain his management responsibility within a corporate structure, the corporation may be managed by board members acting on behalf of the owners. The corporation is subject to several state laws. However this external control of the states is often insignificant as to the operation of the firm.

Corporations are licensed by the state in which corporate structure is sought. Licensing procedures vary from state to state. For example, whereas a single individual can form a corporation in Illinois, other states require two or more incorporates. In addition, some states have rigorous requirements whereas other states are quite liberal as to fees and regulations. The state of Delaware has long been known to bend over backwards to businesses that incorporate. In addition to state laws governing corporations, certain types and sizes of corporations come within the jurisdiction of the Securities Exchange Commission (S.E.C.). Among other requirements, corporations governed by the S.E.C. must openly publish their financial statements.

In addition to a small license fee, costs of incorporation include legal costs that may run from $100 to $1,000 depending on the state and the size of the business in question.

The motivating force behind most proprietors changing to a corporation legal structure is a potential reduction in income tax expense. When one considers the fact that the income tax expense of doing business may exceed $40% of pretax income, even the slightest percentage reduction in tax expense can be a significant dollar savings. While it is not the purpose of this chapter to fully discuss taxation, it is meaningful in the comparison of corporate structure to other legal structures to note a few different income tax implications. As discussed previously, the income tax expense for a proprietorship or partnership is funneled through the tax returns of the individual owners. On the other hand, the corporation pays income tax as a separate entity. In addition, income that is distributed to owners is again taxed to each owner as an individual tax expense.

At first glance, this double taxation for the corporation may seem to be unfavorable and in fact it may be. However, two points are noteworthy before judgment is passed. For one, the income tax rate differs for the proprietorship and the corporation. For relatively small taxable incomes, the tax rate is less for the individual versus the corporation. As the taxable income increases, the effective tax rate becomes smaller for the corporation versus the individual.

A second point has to be recognized in the comparison of the tax liabilities of the corporation versus the proprietorship or partnership. The company owner may be viewed as being double taxed as a corporation; once as an individual and once as the corporation. The corporation can shelter income from double taxation by retaining the income for corporate growth and operations and thus in effect pay a smaller tax rate than it would had it been a proprietorship or a partnership. Secondly, if the income is paid to the owner as salary, it is only taxed once. The only income that is taxed twice is the profits that are paid out the owners. This can be avoided by paying out profits as a bonus (the bonus is taxed as salary).

In addition to different tax rates that relate to a corporate legal structure versus a proprietorship or partnership, the tax "formula"including what is deductible from revenues in the calculation of taxable income differs for the alternative legal structures. These differences and a more rigorous treatment of the determination of tax liability for alternative legal structures of the firm are not discussed here.

A long-standing argument for choosing the corporate structure is centered around the issue of ownership liability. As noted previously, the owners of a proprietorship or partnership have virtually unlimited liability for the debts of the firm. This often proves unfavorable in that should the firm fall on hard times, the owners may loose their house, car, and other personal assets. On the other hand, the liability of the owners (i.e., the

stockholders) of the corporation is normally limited to their initial investment. In addition, they may be liable for the difference between the stated
face value of their ownership stock and the value for which the stock was
issued should the stated face value be greater than the cost. However, stock
is seldom issued below its face value removing this potential liability.

In the case of the construction related firm this limited liability as
to the corporate legal structure is often overemphasized. Due in part to
the relatively high financial risk associated with a construction firm, few
creditors are willing to lend money to construction firms unless the loan is
secured. If the construction firm is a corporation, the creditor will likely require the owners to secure the loan by pledging their personal assets. Thus,
the limited liability of the corporate legal structure loses its significance.

Less significant in importance but also relevant to the decision to incorporate are the considerations of fringe benefits for company personnel and
the ability to raise cash for the firm. Pension, annuities and profit sharing
programs for individuals and businesses are governed by governmental and
internal revenue guidelines. In the case of individuals (i.e., proprietorships
and partnerships) the guidelines are rather restrictive. On the other hand,
liberal pension and annuity programs in addition to stock option plans are
often available to corporate owners and personnel.

Although it is difficult to generalize as to the availability of cash to
alternative legal structures, the corporate structure does provide the best
mechanism for raising cash. Corporations tend to be larger firms with more
assets which they can use to secure loans. Thus, the larger size of the
corporation distorts a comparison of legal structures as to ease of raising
funds. The large firm does not always have an advantage over the smaller
firm. However, it can be stated that large creditor's relatively favorable
attitude towards the corporate legal structure results in the corporation
being in a favorable position as to raising equity capital or issuing debts.
However, being realistic, the contracting firm remains risky in the yes of
investors or creditors regardless of whether the firm is a proprietorship,
partnership, or corporation.

A subchapter S corporation (sometimes referred to as a tax option corporation) is actually a form of a corporation. The subchapter S corporation
offers many of the benefits of the corporation and in addition can be used to
avoid double taxation. Rather than the firm being viewed as a separate entity as to the tax liability, both distributed and non-distributed income can
in effect be funneled for tax purposes to the individual owners of the firm.
The internal revenue imposes several guidelines on the subchapter S corporation as to its income distributions. For example, any income distributed
from the firm's retained earnings in the first two-and-one-half months of
its fiscal year is considered income earned and taxed in a prior year. Such
rules are aimed at eliminating possible manipulation of tax liability.

In order to be eligible to be a subchapter S corporation, a firm must meet certain guidelines that essentially restrict the subchapter S classification to firms that have relatively few owners and have relatively small size. However, there are few constraints as to the dollar value of work the firm performs. The fact that most contracting firms are small firms results in the subchapter S corporation structure being readily available and often advantageous to the construction firm.

## 2.4    Organizational Structure

The basic task of organizing is performed to amass and arrange all required resources, including people, such that the objectives and required work of the firm can be accomplished effectively. Organizing is primarily a people problem. The need for organizing is created because work to be done is too much for one person to handle. Thus it follows, that as a firm grows in regard to its workload, the need and complexity of organizing increases.

The product of organizing is an organization structure. The organization structure determines the flow of interactions within the organization. It determines who decides what, who tells whom, who responds, and who performs what work. If the organization structure is effective it should accomplish the following.

1. Aid coordination
2. Expedite control
3. Emphasize human relations
4. Provide benefits of specialization
5. Pinpoint responsibility

Coordination is a fundamental requirement of an effective organization structure. Coordination enhances communication. Unless procedures, orders, and objectives can be communicated through coordination, individuals will perform their various functions in a less than optimal manner. The running of any firm, including a contracting firm, is a team effort. Very few teams can operate successfully without the coordination of the team members. An estimator cannot accurately price a work item unless he has the aid of past project data from a field supervisor. A construction planner needs the aid of an expediter in determining a feasible schedule for a project. Viewing each function of the firm as being isolated from others defeats effective coordination.

Whereas planning provides the potential for a profitable operation, control is the mechanism by which profits are realized. As such, ignorance of the control function in the organization structure eliminates the potential for an effective organization structure. Much of effective control centers around operative cost accounting and its relation to various individuals and functions in the firm.

An organization structure should focus on the long term as well as the immediate future. An organization structure that fails to recognize and promote human relations is normally short lived in regard to its effectiveness. Failure to recognize "people problems" results in worker resentment, poor worker morale, low worker productivity, embezzlement and theft, and high worker turnover. A people oriented organization structure can facilitate personal management effectiveness.

While assignment of work functions to specific individuals is aimed at overall coordination, a secondary benefit should be higher productivity through specialization. One of the distinct advantages that the large contractor has over its smaller competitor is that the individuals within the large firm can specialize as to their work functions. Whereas the single owner-employee of a contracting firm may have to keep the books, find work, and manage the work; a single individual or group of individuals may be responsible for estimating, another for accounting, another for finance, another for material procurement, another for project management, etc.

The benefit of specialization is that it provides the potential for learning, which can lead to added productivity and quality in regard to the work function. There are potential difficulties associated with specialization that have to be weighed against potential benefits. Overspecialization sometimes leads to isolation and lack of communication. In addition, worker specialization can lead to worker dissatisfaction. In fact, some firms have found that an effective personnel management technique is to rotate workers among work assignments. Other firms tend to have production people specialize, whereas trying to expose management personnel to variable work tasks as a means of personnel development and motivation. The end result is that while specialization can prove beneficial, a fine line exists between effective specialization and inefficient specialization. Thus the difficulty of structuring an organization structure.

An organization structure can provide for two extreme types of decision making. Centralized decision making focuses on decision making by an individual or small group of individuals. Other organizational structures are aimed at decentralized decision making that focus on decision making by groups with each member of the group having somewhat equal contribution in the process. Centralized decision making is characteristic of small firms that are individually owned. On the other hand, the large amount and varying types of expertise that are part of a large firm are best utilized through a decentralized decision making process.

The point to be made is that — regardless of whether a centralized or decentralized process is emphasized in the organization structure — the structure should enable the pinpointing of responsibility for operations, planning, control, etc. The pinpointing of responsibility is necessary if good performance is to be awarded, poor performance corrected, and management objectives evaluated.

The three basic functions of sales, production, and finance have to be carried out by almost every type of profit oriented firm. As the firm grows in size the required functions grow in number and those in existence are often divided into several functions. For example, the basic function of finance is divided into bookkeeping and finance. In the case of a contracting firm it may divide the production function into project estimating and project management.

The functions of the firm and relationships of the functions within the organization structure are illustrated by means of an organization chart. Such a chart is drawn to help visualize what activities are performed and by whom, the work grouping of activities, and their relationships. An organization chart and structure may appear as a chain (sometimes referred to as a hierarchy) or appear circular (alternatively referred to as organic). The two alternatives are shown in Figure 2.1.

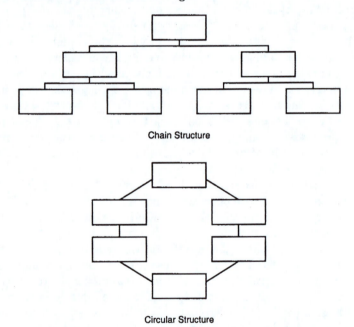

Chain Structure

Circular Structure

**Figure 2.1.** Alternative organizational structures.

Chart lines joining the organization functions indicate the normal flow of communication and decision making where each individual or function is subordinate to that above it. Communication typically proceeds from the top down although the chain structure does at least provide the potential for communication from bottom upwards. The circular structure emphasizes decentralized decision making. The channels of communication of the two alternative structures are summarized in Figure 2.2. Characteristics and utilization of the two organization structure as they relate to the construction firm are discussed in the following two sections.

| Communication Characteristic | Chain | Circular |
|---|---|---|
| Speed | Fast | Slow |
| Accuracy | Good | Poor |
| Stability of leadership position | Firm | None |
| Average morale | Low | High |
| Flexibility to problem change | Low | High |

**Figure 2.2.** Communication characteristics.

# 2.5 The Construction Hierarchical Organization Structure

More often than not, the organizational structure of the contracting firm appears as a chain structure (i.e., hierarchy). While the structure increases in complexity as the firm grows in size, the chain structure remains characteristic of the firm.

The smallest of construction firms is made up of a single individual. The individual assumes all of the management duties of the firm. This person is responsible for sales or marketing, finance and accounting, and production, which entails project estimating, planning, and project supervision. The restricted time considerations of the sole manager typically limits the annual volume of work of the one man operation to a million dollars or less.

As the firm grows in regard to work performed it is necessary to employ more personnel to carry out the management functions of the firm. The owner of the firm is primarily responsible for the sales and marketing function. The amount of financial paperwork of the firm necessitates the employment of an individual to handle the financial concerns of the firm. At this level of organizational size, the finance function is typically characterized by a high degree of bookkeeping with little time spent on financial analysis. More often than not, two individuals will be employed to carry out the production function of this relatively small firm. In particular, one will be responsible for project estimating with perhaps additional responsibilities of material procurement and cost analysis. Yet another production oriented individual will be given project supervision responsibilities and be singled out as a project superintendent. As is true of the estimator, he likely will be responsible for several on-going projects at the same point in time.

Functions within the organization structure are classified as being line or staff functions. Whereas line functions are directly related to the production of the firm's product such as the construction project, staff functions are only indirectly related to production in that such functions support line functions. This is not to say that support functions should be viewed with less importance than line functions. The fact of the matter is that such functions may prove more vital to the accomplishment of company goals than do some line functions. Perhaps the best example of a vital staff function in the organization structure is that of accounting. While accounting personnel do no physical construction, the cost data they provide to line personnel such as project estimators and project superintendents provide these latter individuals the potential to build the project in an efficient manner that controls project time and cost.

The chain or hierarchy type of organization structure is characterized as pinpointing responsibility and authority. Things get done. Decisions are typically made by a single individual or individuals and the resulting procedures and policies funneled down to subordinates vertically. On the other hand, functions tend to become isolated horizontally from one another within such a structure. Information is not totally passed from one function to another, and as such, is not input to many important decisions that are made within the firm. This unfavorable isolation of information and centralizing of decision making is especially noteworthy in regard to the contracting firm. The management and building process that are characteristic of the construction industry results in a need for the recognition of the strong relationships that exist between the management functions. The traditional chain organization structure of the contracting firm is not totally compatible with the recognition of the relationship of the management functions and as such is often less than optimal as to facilitating effective management. The problem can be somewhat alleviated by providing links of communication between dependent functions and promoting the usage of such links.

## 2.6   Organic Structure for the Construction Firm

The organic or circular organization structure is aimed at a sharing of information, ideas, and feedback. It is based on the theory that there is a set of information and data that are common to decision making as it relates to several management functions. The sharing of the information and data are to facilitate optimal decisions and provide an efficient and effective organization structure.

Much of the management of the construction firm relates to project management. One of the strongest relationships between management functions is evident in project management. The key to the relationship of the functions is a system by which information system based on past project data. As projects are performed, cost and productivity data are funneled into the accounting functions where the data is structured for use by the payroll function, project supervision function, the estimating function, the planning and procurement functions, and further use by the accounting function. Each of these functions are related and relevant information is funneled back and forth to the individuals and departments responsible for the functions. For example, cost and productivity data are funneled from past projects to the estimating individual in charge of pricing a future project. The completed estimate serves as the basis for the work carried out through the project planning and procurement functions. In addition, past project data are quickly structured and analyzed and compared to the project plan and estimate. The project superintendent uses this comparison as the basis of project control.

The point to be made is that a properly designed information system that enhances communication between the project management functions results in each function aiding the optimization of another. In order for such an information system to be effective, the organizational structure has to be compatible with the system. The organic or circular organizational structure has the potential for such compatibility.

An organic organization structure is not without potential difficulties. The success of the structure is dependent on the total cooperation of each individual that is part of the structure. In addition, when one decentralizes decision making — as is a characteristic of an organic structure — the potential for confusion and lack of strong leadership and decisive decision making increase. On the other hand, given the cooperation of all, not only is a better and more complete decision enabled, but morale tends to improve in that individuals and groups within the organizational structure become part of a team effort.

## 2.7 Financial Structure: Leverage and Liquidity

The success of a profit oriented firm is often measured by the firm's earning power. The earning power is the ratio of the firm's net operating income divided by its net operating assets. In effect, it is a measure of the firm's ability to maximize its return on its investment.

$$\text{Earning power} = \text{Margin} \times \text{Turnover}$$
or
$$\frac{\text{New operating income}}{\text{Net operating assets}} = \frac{\text{Net operating income}}{\text{Revenue}} \times \frac{\text{Revenue}}{\text{Net operating assets}}$$

The firm can increase its earning power or its return on investment through operations. This is evident from inspection of the formula for earning power.

As to the contracting firm, the margin is merely the ratio of project profit to project contract value. The turnover relates to the volume of work the contracting firm performs versus the dollar value of assets of the firm. By means of increasing their volume of work performed, reducing their expenses, or decreasing their assets, the earning power of the firm can be increased. Unfortunately, this increasing or decreasing of factors that influence earning power does not linearly affect earning power. For example, if the firm decreases its invested assets, it will likely result in less volume of work. In addition, increased volume of work may add to overhead expenses and thus lower the profit margin. The end result is that the increasing of earning power through variations in operations is complex and the output of a given management adjustment may not be known until after the fact.

An equally important aspect of maximizing a firm's rate of return on its investment relates to the financial structure of the firm. In particular, the assets of the firm can be obtained through issuance of debt to creditors or by means of investment of equity capital by the owners of the firm. This invested capital is sometimes referred to as owner's equity, or residual owner's equity. In the following discussion the terms debt and equity will be used to represent the two segments of the capital aspect of the financial structure of the firm.

The previous discussion of earning power assumed that the assets of the firm were in fact obtained through equity investments. However, the firm may in fact be able to generate profits through the purchase of assets that are obtained by the issuance of debt. Assuming that the interest rate associated with such debts is favorable, the resulting earning power will be increased. For example, let us assume for simplicity purposes, that $10 of assets are purchased through issuance of $2 of debt and investment equity of $8. Should the firm earn $1, the rate of return (i.e., the non-adjusted earning power) on total assets is $1/10 or 10%. On the other hand, realizing that only $8 has been invested (i.e., the net assets), the true earning power (i.e., the return on equity) is $1/\$8 or 12.5%. Thus, the earning power is in fact increased due to the financial structure. The difference between the earning power of 10% and the return on equity of 12.5% is produced by financial leverage. If we let the return on equity be designated as $R$, earning power (i.e., non-adjusted) as $EP$, and the leverage factor as $L$ then

$$R = EP \times L \quad \text{or} \quad L = R/EP = 12.5/10 \quad \text{or} \quad 1.25$$

Thus the earning power is magnified by 1.25 to produce the rate of return on investment capital.

The above calculation of the leverage factor ignored the effect of interest which the owners of the firm have to pay creditors when issuing debt. The presence of a high interest cost may in fact negate the positive effects of earnings obtained through the issuance of debt. Looked at another way, the leverage factor $L$ represents the ratio of the earnings actually received by the residual owners to the amount they should have received based solely on their proportionate contribution of funds to the business. Let $T$ equal the proportion of total assets financed by the investment equity (i.e., 80% in the previous example) and $Y$ equal the total earnings. The earnings the owners should receive based on their proportionate share is $TY$. However, they also receive earnings due to the issuance of debt. The additional earnings they receive is equal to the difference between what the creditor's share would be (i.e., $Y - TY$) and what the creditors actually receive in interest $I$, where $I$ is the annual interest paid to the creditors. Thus, the bonus earnings to the owner are

$$(Y - TY) - I \quad \text{or} \quad (1 - T)Y - I$$

As long as $(1 - T)Y$ is greater than $I$, the leverage is favorable. The actual leverage factor, considering the interest effect, can be calculated as follows:

$$\text{Leverage} = \frac{\text{Owner's proportionate earnings} + \text{Bonus earnings from debt}}{\text{Owner's proportionate earnings}}$$

$$\text{Leverage} = \frac{TY + ((1 - T)Y - I)}{TY} \quad \text{or} \quad \frac{1(Y - I)}{T(Y)}$$

Assuming an annual interest charge of 5% in the previous example, the leverage would be calculated as follows.

$$L = \frac{.80(\$1) + ((1 - .80)(\$1) - \$.10)}{.80(\$1)} = \frac{.80 + .10}{.80} = 1.125$$

This would result in a 12.5% increase in earning power. Note that as long as the annual interest charge is less than $.20 (i.e., 10% interest rate on the $2.00 debt) the leverage is favorable in that the leverage factor is greater than one. If the interest charge is $.30 the unfavorable leverage factor would be as follows.

$$L = \frac{.80 - .10}{.80} = .875$$

This would result in a decreasing of the unadjusted earning power of the firm. The end result is that leverage tends to multiply a positive earning power or negative earning power.

One other consideration can result in a slight modification in the determination of the leverage factor. Interest is a deductible expense to the firm and thus a benefit of issuing debt. This added benefit can be recognized in the leverage formula. However, because the modification is slight as to the leverage factor it will be ignored here. Ignoring any tax effects and assuming a given income, the two factors that influence the determination of the leverage are interest and the percentage ownership by equity versus creditor debt. Assuming a favorable leverage factor (i.e., greater than one), the smaller the interest charge, the more favorable the leverage factor. Naturally, a zero interest charge would be ideal.

From inspection of the formula for the leverage factor, it can be observed that the percentage ownership of the firm through investment equity also effects the determination of the value of the leverage. Assuming a favorable leverage, the smaller the proportionate investment ownership of the assets the more favorable the leverage factor. In fact, the determination of the value of the leverage is much more sensitive to the ownership percentage than it is to the interest charge.

The question might be raised, "Why not decrease the proportionate equity ownership of the firm relative to the debt ownership?" Theoretically, the proportionate equity ownership can be reduced to a percentage that approaches zero (i.e., admittedly it would be difficult to continue to get creditors to increase their lending percentage). From the previous discussion it follows that such a decrease in equity ownership would provide substantial earning power benefits to the owners. However, it is noteworthy that only the issue of return to the owners has been raised in the discussion of leverage and financial structure. An important factor that has been neglected is the issue of risk.

Two somewhat independent points should be recognized when considering the leverage and financial structure of the firm as they relate to risk. In order to obtain leverage, debt has to be issued and this debt implies an interest charge. An interest expense is a fixed cost to the borrower of money. It comes due regardless of the amount of income obtained through operations. It can create a liquidity problem for the firm in that it reduces the amount of liquid assets (cash or assets readily transferable to cash) available to the firm. In many cases, bankruptcy problems start with the unavailability of cash and other liquid assets such as receivables. Ideally, a firm that is high leveraged should have high liquidity such that they are in a position to pay interest when it comes due.

A second relevant consideration of leverage and risk relates to possible decreases in income. As income decreases, a favorable leverage factor greater than one can decrease to an unfavorable value less than one. This is evident from the fact that as the value of $Y$ decreases, $((1 - T)Y - I)$ becomes smaller and can become negative, thus resulting in a leverage factor less than one. In addition, the leveraging effects tend to multiply the negative effects as income continues to decrease. The end result is that return has to be balanced with risk when considering leverage and the financial structure of the firm. Consideration of expected income, interest expense, liquidity, and to a lesser degree taxes, are necessary to the financial structure decision. The uncertainty of some of these factors such as income, increase the complexity of the analysis. However, recognition of the relevant factors can go a long way in generating profits and decreasing profits through financial structuring of the firm.

## 2.8   Financial Structure of the Construction Firm

Leverage and liquidity are two important aspects of the financial structure considerations that relate to the financial soundness, income, and growth of the contracting firm. In addition the actual makeup of the balance sheet items of the firm related to the overall strength of the firm.

Unique external factors tend to shape the total financial structure of their firm. The inability to raise equity capital, the operating on short-term credit, the large amounts of receivables on the books because of the owner's retention of part payment, and the dependence of some firms on large dollar values of equipment all dictate the financial structure of the construction firm. Some of these industry characteristics result in a less than favorable structure that contains an undue amount of risk for the firm.

The primary assets of any firm are its cash, receivables, inventory, and fixed and other assets. In comparison to all other industries, the contracting firm has a high percentage of its assets in receivables. In addition, compared to the manufacturing industry (of which construction is a part), the construction firm has a relatively small percent of fixed assets.

The contractor firm's high dependence on receivables results from the billing and payment retention procedures that are characteristic of the industry. Upon completing various work items, the construction firm bills the owner. However, actual payment may be slow depending on who is the owner. Delays of a month or more are common. In addition, when in fact the payment is made, the owner may choose to retain part of the payment (typically 10%) until the construction project is substantially completed. Considering the fact that projects may have a contract value exceeding $100,000, this 10% retention can amount to a substantial dollar value receivable.

Receivables are troublesome assets. For one, they do not provide the contracting firm an asset that can be used for operations and growth. It is true that the firm can convert them to cash by means of assignment or factoring, however such procedures are not readily available to the construction firm and when they are they are relatively unfavorable as to interest cost. As such, receivables typically do not aid the firm in resolving any cash flow problems. In fact, a large buildup of receivables is often characteristic of a cash flow problem leading to bankruptcy. The uncertainty and risk associated with receivables are other unfavorable financial considerations. Many a construction firm has had to enter court in order to collect its receivables.

The contracting firm has a relatively large percentage of its assets in cash. The construction firm continually has to have cash on hand to pay for varying amounts of labor and material expenses. While an adequate amount of cash for this purpose is necessary in order to avoid cash flow problems, the fact remains that cash is not a good income producing asset. Cash heavy firms tend to be non-growth firms in that income and growth are generated by means of investment in non-monetary assets such as plant and equipment. The construction firm has to be cautious as to using its cash for non-monetary investments due to its need to have a large amount of cash readily available for operations.

As to the capital side of the construction firm's balance sheet, the contracting firm is often characterized by a high degree of debt. Whereas typical manufacturing firms have approximately the same amount of debt capital and equity capital (i.e., invested capital plus retained earnings), the contracting firm typically has liabilities in excess of their equity capital. This higher dependence on debt is not unexpected when one considers the difficulty that the construction firm has with raising investment capital. Typically the construction firm is relatively small and not in a favorable position to issue stock that provides equity capital.

A large percentage of the contractor's liabilities are current liabilities. Normally this means that they are liabilities that must be repaid within a duration of a year versus long-term liabilities that typically can be repaid in periods of ten or twenty years or more. The lack of significant long term liabilities relates to several factors. For one, similar to investment capital, third parties viewing the contracting firm with risk are reluctant to commit funds to the firm on a long-term basis. Secondly, many contracting firms lease their construction equipment by means of a long term contract.

The contracting firm's high dependence on current liabilities places it in a position of high financial risk. By definition, current liabilities come due in a relatively short period of time. Should the firm have difficulty making payment due to a slump in its operations or poor cash flow planning, it in effect defaults and becomes subject to bankruptcy proceedings. In order to operate and to reduce some of the financial risk of current liabilities, the contracting firm should seek a sound and flexible credit line with a lending institution.

## 2.9  Summary

The ease of entrance into the construction industry has resulted in numerous new entrants annually. More often than not, little thought and effort is used in the structuring of the firm. Instead entrants "play it by ear" with the result being that the structure of the firm is shaped by the environment of the firm. The end result is a structure that is not optimal as to legal consequences, liabilities, tax benefits, organizational operation, and financial soundness and stability.

The first step on the road to a long lasting financial successful construction firm is a thorough analysis and selection of the type of work the firm is to perform; and its legal, organization, and financial structure. The decision to be a general contractor, construction manager, proprietorship, partnership, etc., is a decision that plays as an important role in generating future profits as other management functions such as personnel management, and project cost control. Equally important is the decision as to the firm's organization structure and financial structure. The organization structure provides the means of efficient management communication, effective personnel management practices, and required company project control.

The construction industry is affected by unique external factors that in part dictate the financial structure of the firm. The inability of the construction firm to raise equity capital, its high dependence on short-term credit, its low investment in raw materials, and the large amount of current receivables on its books owing to the process by which it receives payment from project owners, all place constraints on the financial structure of the firm. Within these constraints the firm has to strive to build a sound financial structure such that it remains solvent, and to enable itself to be in a favorable position in its dealings with potential creditors and investors.

# Chapter 3

# Marketing: Obtaining and Maintaining Clients

## 3.1 Introduction

It is common to think of marketing as selling. If one accepts this definition, the contractor's involvement with marketing is minimal in that the firm does not have to "sell" its product in that the firm typically either wins or loses a project contract in the competitive bidding process. There are two fallacies with this conclusion that one might reach. For one, negotiated contracts — whereby the construction firm obtains work on the basis of its promotional efforts — its ability to perform and its overall "selling" efforts, are becoming increasingly common in the construction process. Secondly, contrary to the beliefs of many who think of marketing as only selling, the study and application of marketing encompasses much more than the selling of one's product or service.

Marketing is the planning, pricing, promoting, and distribution (also referred to as placing) of one's product or service. Thus, reference to the four Ps (planning, pricing, promoting, and placing) is common when a broad and complete definition of marketing is given. The selling function that was referred to earlier is more properly referred to as the promoting aspect of marketing.

The contractor, similar to an automobile manufacturer, produces a product that it must market to a purchaser. Although the construction firm's marketing of its product differs from that of other manufacturers, such as the automobile manufacturer, the construction firm has to perform the same functions of planning, pricing, promoting, and placing. As will be discussed in this chapter, the firm's emphasis on each of the four marketing tasks may vary from that of other producers of manufactured goods. This change of emphasis is a result of the differences of the environment of the construction industry from that of other industries.

It has been correctly stated that the marketing business function is fundamental to the necessity of all other business functions. That is, there would be no need for such business functions as business law, financing, or personnel management unless a product or service is marketed by the firm. With this in mind, this chapter focuses on marketing terminology and practices as they relate to the contractor.

## 3.2 Marketing Environment of the Construction Firm

Required marketing practices differ for the contractor depending on whether the firm engages in competitive-bid contracts or negotiated contracts. In the past, negotiated contracts were limited to small construction projects. However, an increasing number of large negotiated contracts are being entered into by the contractor. The end result is that many firms are finding themselves entering into both types of contract agreements.

The firm is basically marketing cost in the competitive-bid process. Much of the firm's efforts center on submitting a bid on a project that has a cost low enough to be the low bid yet large enough to provide the firm a profit. On the other hand, the firm seeking negotiated contracts, while concerned with cost, has to pay added attention to external relations. As such, special care must be directed to turning out quality work and keeping and obtaining personal contacts with potential owners through associations, social clubs, and in some cases providing free work or donations to charitable organization.

As to its type of operation, most business services and the government classify the contractor as a manufacturing firm. However, in reality the firm is somewhat of a combination of a manufacturer and a service oriented firm. It is true that the traditional firm manufacturers a construction project. However the project is not truly the firm's product in that the project is built for an owner and the project has in fact been designed by the owner through his architect/engineer. In one sense the firm is providing a service function in that it is providing management and technical skills to the owner such that the owner's project can become reality.

In that the contractor is not normally responsible for the conception of a project, its marketing tactics do not have to concentrate on customers' needs and satisfaction and pricing practices such as markups, markdowns, etc. On the other hand, because the firm has some of the characteristics of both manufacturing and service industries, its marketing tactics have to recognize both physical commodities such as concrete and service attributes such as ethics and professional external relations. The latter attributes of

ethics and professional external relations are becoming increasingly important with the increasing popularity of negotiated contracts, design-built,
and construction management contracts.

Most of the unique marketing aspects of the firm relate to the fact that
each project the firm builds is somewhat unique. Much of the marketing
study, research, and publications relate to the marketing of a product that
is produced and sold several times. Because this is not a characteristic of
the construction product, many of the tradition manufacturing principals
are not applicable to the construction industry. This fact complicates the
study of planning, pricing, promoting, and placing as it relates to the firm
but it does not follow that the firm can pass over these marketing functions.

## 3.3   Planning for Company Operations

The first step in a successful marketing process is that of planning. By definition, planning is the selection of objectives and their means of attainment.
As it relates to contractor marketing, planning addresses the determination
of the type of work the firm is to perform, the decision as to where it is
to perform work, the determination of the organizational structure for the
marketing of the firm's product (i.e., the construction project) such that
the firm can price, promote, and place the product with a client in a manner
that satisfied the firm's profit oriented objectives. This later planning task
entails many of the traditional construction management functions such as
project method analysis and improvement, and overall project scheduling.
Yet another planning oriented function, one that is often overlooked, is
the determination of growth objectives and the corresponding policies and
procedures.

The three basic business functions are defined as sales or marketing,
production, and finance. The owner of the small or medium sized construction firm performs the marketing function himself. However as a firm grows
in size, several individuals are required to carry out the function. In fact
the owner or president may see it appropriate to remove himself from the
function entirely in favor of setting up a distinct marketing department as
shown in Figure 3.1. In fact several large contractors have created two marketing related departments, one referred to as marketing and one referred
to as sales. While not typical of the contracting firm, some businesses will
expand the marketing function to yet a third department referred to as
research.

As is true of all management functions, the marketing function should
not be isolated from other functions. The organizational structure should
be such that it facilitates communication between marketing, production,
and finance. For example, past project data from the production function
provides planning data for marketing. In addition, in order to properly fi-

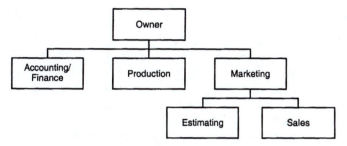

**Figure 3.1.** Organizational structure.

nance operations, the finance function must have marketing plans available for evaluating alternatives. Examples of additional functional relationships are numerous.

The newly formed contractor often chooses its area of work concentration on the basis of the skills of the founders. While this seems logical, it also is true that more attention should be given to the market for the type of work being considered. Many a contractor has commenced operations with a focus in his home city only to later find that the demand for his type of work is short lived.

The point to be made is that the contractor should have knowledge as to the potential construction expenditures in its area of specialty. While the dollars spent in different categories vary somewhat from one year to the next, substantial changes are unlikely. Various government services keep records as to various types of construction expenditures.

Perhaps more important than the concern for the breakdown of total expected construction expenditures, is the concern for expected expenditures as a function of geographic locations. The fact that residential expenditures are 36% of all construction expenditures is not totally relevant to a construction firm doing work in a town where residential work is only 5% of construction expenditures in the community. Variations in the makeup of construction expenditures occur from one part of the country to the next due to environmental differences, variations in local economics, population growth patterns, and to a lesser degree material availability.

The process by which the firm seeks to determine potential construction expenditures in a given geographic location is referred to as marketing research. Marketing research can embrace several somewhat independent activities such as the following.

**Market Analysis:** the study of the size, location, and characteristic of markets.

**Volume Analysis:** an analysis of past sales such as volume and type of construction

**Consumer Research:** concerned with the analysis of consumer attitudes, reactions, and preferences.

By carrying out one or more of these marketing research activities, the contractor can take out some of the guesswork as to expected construction expenditures in a given location. While it is unrealistic to think that the typical contractor will or can engage in extensive personal surveys, less extensive marketing research can be carried out with little effort or expense. For example, by merely consulting published governmental data, existing and projected population growths in various geographic locations can be determined.

More often than not, geographic factors (i.e., hometown considerations) dictate the location of the contractor. However, marketing research can still prove useful to the firm fixed to a given location. Observation and analysis of various markets and common attitudes within a given location can lead the contractor into new markets such as entrance into residential work or heavy and highway work. Analysis of planned public construction expenditures, analysis and projections of industrial capital expenditures, recognition of current and projected interest rates (i.e., residential construction being very sensitive to interest rates), and an awareness of existing and expected competition from other contractor firms in the geographic area are all worthwhile marketing research practices. Daily reading of a business publication such as the *Wall Street Journal* or local financial institution economic publications are other means of carrying out marketing research.

The firm's marketing plans have to recognize the firm's objectives as to growth. Different pricing and promoting procedures and policies are justified given different growth objectives. For example, a growth objective aimed at increasing company work load may justify lowering of the firm's profit margin such that it can be more competitive as to obtaining new work. In addition, such an objective may justify an escalated advertising effort, and the added overhead expense of branch offices. On the other hand, should the firm have a growth objective that centers on a constant work volume but an increased profit margin, different pricing and promoting procedures and policies may follow. For example, emphasis on productivity and obtaining clients more interested in quality than low cost may be part of the firms' pricing and promoting policies.

The planning aspect of marketing can also be thought of to include the more specialized construction project management practices of project planning, scheduling, material procurement, etc. These practices along with project estimating border between the planning and pricing aspects of marketing. While emphasis is not given here to project management tools such as networks, optimization, work analysis and improvement, etc., these tools are important elements of marketing planning. For example, the ability to utilize work improvement techniques enables the construction firm to competitively price certain types of work which in part dictate the type and amount of work the firm undertakes.

## 3.4   Pricing the Firm's Work

Pricing is considered by many to be the key activity within the capitalistic system of free enterprise. The pricing aspect of marketing dictates both the profit margin of the firm and its volume of work performed. As to the construction industry, pricing takes on added importance. Construction contract specifications almost totally dictate the degree of quality the construction firm is to attain when constructing the construction project. As such, quality of work somewhat loses its significance in regard to obtaining new work. This is especially the case when the competitive bidding process is used to secure a construction firm.

Typically, contractor is basically selling price to the construction project owner. Assuming quality is determined by the owner through his architect/engineer, time is the only other variable the project owner is in effect purchasing. However, time to construct projects can be expressed monetarily. That is, time is money in that rent is lost when a project remains unfinished, fees are lost when a recreational facility is not ready for use, etc. The end result is that the contractor firm's ability to serve up the construction project to an owner within budgeted cost and time is the best marketing technique available to the contractor. Naturally such a cost should provide the contractor firm with a reasonable profit.

Because of the contractor firm's type of product, it is in a rather unique position as to pricing. As noted earlier, the contractor firm can be categorized as being somewhere between a manufacturing firm and a service oriented firm. Sales volume pricing theories for manufacturing firms do not totally hold true for the construction firm in that the firm does not sell repetitive types of units that are subject to demand price sensitivity factors. On the other hand, like other contractor firms, the construction firm has fixed expenses and variable expenses that vary with the volume of work performed.

In the sense that the firm contracts with a client to build the cleint's project, the contractor firm is providing a service. The pricing theories of such service businesses focus on expected volume of work performed. That is, fixed expenses of service industries are often considered to be insignificant in pricing procedures and policies.

The business environment of a service firm or industry is often characterized as being perfect competition or monopoly. In reality no industry finds itself in either of the two extreme environments. However, the construction industry approaches perfect competition more closely than any other industry. This is especially the case in large cities where several firms participate in the construction industry. The presence of several firms prevents any one firm from dictating the contract price and profit margin. Rather the contract price and profit margin result from the competition of all firms active in the geographic area.

Even in an area of limited population, the fact that firms outside the geographic come in to bid work results in somewhat perfect competition. The number of geographic areas where a single firm or a very few number of firms bid construction work is decreasing. When in fact a firm does find itself in the position of being able to dictate the contract price it is characterized as operating in a monopoly.

The large contractor firm competing against much smaller firms may sometimes be able to use price strategy characteristic of a firm operating in a monopoly. The large firm's ability to lower its profit margin, or its smaller percentage overhead, may allow them to dictate project contract price. The large firm may settle for little or no profit in the short run with the objective of eliminating long-term competition from other firms.

The ease of entrance into the construction industry, the mobility of the contractor firm, and the bidding process which is used in the industry typically results in policies based on a monopoly being short lived. The end result is that regardless of whether the contractor firm obtains work through competitive bidding or negotiated contracts, its pricing procedures and policies will eventually have to recognize competition.

## 3.5   Pricing Parameters

Competition is one of several variables the contractor has to recognize in its pricing of construction projects. The variables to be recognized can be classified as long-term or short-term considerations. Long-term considerations or variables include the following.

1. Historical pricing practices and profit margins of the industry.
2. Price required to generate capital for growth.
3. Price and profit margin required to facilitate raising of debt and equity capital.

These and perhaps other long-term considerations should shape the overall pricing aspect of marketing for the contractor firm. For example, it is unrealistic to think that a given contractor firm's pricing practices can completely ignore the profit margin of the overall construction industry. The near perfect competition characteristic of the industry prevents a single firm from isolating itself as to the marketplace.

Short-term variables may cause a need for short-term adjustments in the pricing procedures and policies of the construction firm. Some of these variables are as follows.

1. State of present and expected future economy
2. Need for work
3. Dollar value of project being considered
4. Duration of project being considered
5. Expected project risk
6. Expected bidding competition on project being considered

The first two variables noted are related. That is, the state of the economy, both nationally and locally, will likely affect the individual contractor firm's need for work, it may be satisfied with a short-term pricing policy that provides for a lessening of the long-term desired profit margin. The same type of policy might be used when the firm foresees a future slowdown in construction activity. Again, in order to cover fixed expenses, taking on work with little or no profit margin included may prove to be a financially sound policy.

The dollar value and duration of a project being considered may also effect the pricing of the construction project. In particular, some contractor firms have a pricing policy of somewhat decreasing their profit margins as the dollar value of a project increases or the expected duration of a project decreases. Such a policy is based on the fact that a smaller profit margin on a large dollar value project (i.e., all other factors being equal) will still yield a considerable total dollar profit. In addition should the duration of a project be short, a small profit margin or total dollar profit will still yield a favorable rate of return due to the short period of time for which the firm ties up its assets.

Expected project risk is yet another short-term project pricing variable. While risk is difficult to measure, surely a firm's pricing procedures should attempt to account for varying amounts of risk that are characteristic of individual projects. Such risk may be in the form of expected weather difficulties, problems in coordination of subcontractors, labor union difficulties, etc. One justifiable pricing policy would be merely to avoid such projects. At the very least, an upwards adjustment in profit margin is needed.

Perhaps of all the short-term variables noted, the one that has received the most attention as to pricing policies is that of expected bidding competition. There is an increasing amount of study being directed towards the modeling of pricing as it relates to expected bidding competition. More likely than not, the increased interest in this area is generated by the fact that a quantitative solution can be formulated for the modeling of pricing as a function of expected competition. The fact remains that all of the short-term variables are important in the pricing aspect of marketing. In fact, if one was forced to weigh the factors as to importance, the need for work would certainly be on top or near the top.

The efforts aimed at structuring the pricing function to recognize expected bidding competition have resulted in a so called bidding strategy model. The bidding strategy model is especially relevant when a firm bids frequently against the same firms. This is usually the case for the contracting firm. Let us now consider the bidding strategy pricing model.

## 3.6   Bid Strategy

The optimal profit that a contractor should add to his bid proposal is partly determined by the contractor competition for the project. This is particularly true in the competitive bidding procedure. The contractor firm's winning or losing a project contract often depends on how well he has formulated his information about his competitors. The contractor must always search for ways to gain an advantage over his competitors. This includes using new and cheaper construction methods and management practices. One of the ways a contractor might gain an advantage over his competitors is to formulate bidding information about their past performances. The contractor can formulate this information into some type of bid strategy. For our purposes, we will consider bid strategy as a combination of various bidding rules a contractor follows for bidding, based on a formulation of information. In our case, the information will be past bids of contractors. Formulating the information is often referred to as determining a bidding model.

The profit in a contractor's bid for a particular project is the amount of money he intends to make on the project. For purposes of discussion, let us assume that the contractor's estimated cost of bidding a project is indeed accurate and is equal to his actual building cost. Therefore, on a particular project for which a contractor submits a bid, the firm will either receive it's desired profit (assuming the firm received the project), or will receive zero profit (assuming the firm does not receive the project). It becomes clear that a contractor's long-term profit (average profit over a long period of time) will not only be a function of the profit within the firm's bids, but a function of how many projects he receives from the number of projects for which he submits bids.

Owing to the possibility of two levels of profit (depending on whether a contractor wins or loses the contract for a project), it is necessary to define two different types of profit. A contractor's immediate profit on a project is defined as the difference between the contractor's bid price for the project and his actual cost of building it. Assume a contractor's bid is equal to $X$. If we let $A$ represent the contractor's actual cost of building the project, then the contractor's immediate profit on the project ($I.P.$) is given by the following formula.

$$I.P. = X - A$$

If a contractor submits a high bid (a large profit included), his chance for receiving the contract in a competitive bidding environment is very small. As he reduces his profit, and therefore his bid, his chance for receiving the contract increases.

If we assign probabilities of receiving the contract to the various bids the contractor considers feasible, we can calculate an expected profit for the various bids. The expected profit of a particular bid on a proposed project is defined as the immediate profit of the bid for the project multiplied by the probability of the bid winning the contract. In a competitive bidding procedure, winning the contract implies that the bid is the lowest responsible bid. If we let $p$ represent the probability of a particular bid winning the contract for a project, then the expected profit of the bid ($E.P.$) is given by the following formula.

$$E.P. = p(X - A) = p(I.P.)$$

Assume a contractor is interested in a certain project, a project the contractor estimates to cost $20,000 to build. The contractor has a choice of submitting 3 different bids for the project. These bids and their probabilities of winning the project contract are as follows.

| Bid name | Amount | Probability of winning contract |
|---|---|---|
| B1 | $30,000 | 0.1 |
| B2 | $25,000 | 0.5 |
| B3 | $22,000 | 0.8 |

The probabilities shown are estimated from the contractor's evaluation of the chance of being the lowest bidder. Of course, bid B, has the highest immediate profit ($30,000 - $20,000,$ or $10,000$). However, because of the low probability of winning the bid, it may not be the best bid to make to maximize overall profits. The calculation of the various bid's expected profits is as follows.

| Bid name | Probability × Immediate profit | Expected profit |
|---|---|---|
| B1 | 0.1 × (10,000) | $1,000 |
| B2 | 0.5 × (5,000) | $2,500 |
| B3 | 0.8 × (2,000) | $1,600 |

It is observed that Bid B has the highest expected profit. Expected profit may be conceived as representing the average profit a contractor can expect to make per project if he were to submit the same bid to a large number of similar projects. Expected profit does not represent the actual profit the contractor expects to make on a project. In the problem described, the contractor would either make a profit of 0 or a profit of $5,000 if he submitted Bid B whereas the expected profit is calculated to be $2,500. Immediate profit does not recognize the probability of a bid winning a contract. Expected profit recognizes the objective of maximizing total long-term profits. Thus expected profit calculations are more meaningful than immediate profit calculations, and should be used to determine the optimal profit and bid. In the example described the contractor should submit Bid B.

By using the discussed expected profit concepts in addition to information concerning the past bids of his competition, a contractor can develop a bidding strategy which he may use to optimize his profits. At a competitive bid letting it is common practice to announce openly all the bids of the respective contractors. The contractor can record the bid prices of the contractors, along with his own bid and his own estimate of the project's cost. If it is possible for him to learn the actual cost of the project (either through building it himself or through information obtained from others), he should also record this information. The contractor should also be aware of any special conditions, such as knowledge about a particular contractor's need for work. Having recorded the past bidding information about his competitors, the contractor should also take note of any special conditions, such as knowledge about the work required for the project. Having recorded this past bidding information about his competitors, the contractor formulates the information about his competitors, and then can formulate the information into a bidding strategy for future projects. Naturally, the more information the contractor has available, and the more accurate his information, the better is his chance of having his strategy prove successful.

When a contractor is bidding on a project in a competitive bidding atmosphere, he generally finds himself in one of the following states regarding his competition. In the most deterministic or ideal state, the contractor knows who each of the competitors will be on an upcoming project. A somewhat infrequent situation occurs when the contractor knows how many competitors there are for the project, but does not know who they are. Since there is less information available in this case than in the case of known competitors, the bidding strategy will be less deterministic and therefore less reliable than the bidding strategy for known competitors. A less deterministic and less desirable situation occurs when the contractor knows neither who the competition is, nor how many competitors there are. The bidding strategy for this situation will be less reliable than the previous two cases, owing to the lack of more complete information.

Consider the case in which the contractor knows who his competitors will be for a competitive bid project letting. In particular, let us assume that the contractor knows he is only going to be competing against one contractor (contractor XYZ). Assume the contractor has bid against contractor XYZ many times in the past and has kept records of contractor XYZ's bids. For each of the projects the contractor has also recorded his estimated cost. Having this information, the contractor can calculate the ratio of contractor XYZ's bid price to the contractor's cost estimate for the various projects. The contractor's recorded information is summarized as follows:

| Contractor XYZ's bid/ Contractor's estimated job cost | Frequency of occurrence |
|:---:|:---:|
| 0.8 | 1 |
| 0.9 | 2 |
| 1.0 | 7 |
| 1.1 | 12 |
| 1.2 | 21 |
| 1.3 | 18 |
| 1.4 | 7 |
| 1.5 | 2 |
| Total | 70 |

Having the frequency table of the various ratios, the contractor can calculate the probability of each bid ratio by dividing each bid's frequency of occurrence by the total number of measured bid occurrences. For example, the probability of a ratio of 1.0 is 7/70, or 0.10. Probabilities of the other ratios are as follows. The probabilities are rounded off to two decimal places.

| Contractor XYZ's bid/ Contractor's estimated job cost | Probability |
|:---:|:---:|
| 0.8 | 0.01 |
| 0.9 | 0.03 |
| 1.0 | 0.10 |
| 1.1 | 0.17 |
| 1.2 | 0.30 |
| 1.3 | 0.26 |
| 1.4 | 0.10 |
| 1.5 | 0.03 |
| Total | 1.00 |

Having calculated the probabilities of the various ratios, the contractor can calculate the probability of his various bids being lower than contractor XYZ's bids. To eliminate theoretical bid ties, it will be assumed that the contractor will bid different ratios than the computed ratios for contractor XYZ. For example, to be lower than contractor XYZ's bid-to-cost ratio of 1.10, the contractor might make a bid with a bid-to-cost ratio of 1.05. Let us assume that the contractor decides upon the following bid-to-estimated-cost ratios as being feasible.

| Contractor's bid/ Contractor's estimated job cost | Probability that contractor's bid is lower than bid of XYZ |
|:---:|:---:|
| 0.75 | 1.00 |
| 0.85 | 0.99 |
| 0.95 | 0.96 |
| 1.05 | 0.86 |
| 1.15 | 0.69 |
| 1.25 | 0.39 |
| 1.35 | 0.13 |
| 1.45 | 0.03 |
| 1.55 | 0.00 |

The calculated probability of being the lowest bidder, or winning the contract, for any particular bid ratio is found by merely summing all the probabilities of contractor XYZ's ratio being higher than the particular bid. For example, if the contractor is to make a bid with a bid-estimated-cost ratio of 1.35, the probability of winning would be the sum of 0.03 (the probability that the ratio of XYZ's bid to the contractor's estimated cost is 1.5) and 0.10 (the probability that the ratio is 1.4). Thus, the probability of a contractor bid with a bid-to-estimated-cost ratio of 1.35 winning the contract is 0.13.

The contractor may now use this information to form a strategy for bidding against contractor XYZ. He may do this by calculating the expected profits of his possible feasible bids. Expected profit of a bid was defined as the immediate profit of the bid multiplied by the bid's probability of winning the project contract. Immediate profit for a bid was defined as the bid price minus the actual cost of the project. Let us assume that the contractor's estimated cost of the project is equal to the actual cost. Ideally the contractor would like his estimator to estimate the actual cost correctly, but this is not always the case. However, owing to the lack of information about the actual cost of the project, let us assume that the estimated cost of the project is the actual cost. The immediate profit for each of the contractor's possible bids then becomes equal to the bid price minus the estimated cost of the project. The bid prices are given in terms of the estimated cost of the project. Letting $c$ equal the estimated cost of the project, the immediate profit of the contractor's possible bids may be stated in terms of $c$. The immediate profit of the bids may be found by merely subtracting $1.0c$ (the estimated cost of the job) from the respective bids. For example, for a bid of $1.35c$, the immediate profit is $1.35c$, the probability of winning against contractor XYZ was calculated as 0.13; therefore, the expected profit is 0.13 multiplied by $0.35c$, or $0.0455c$. The expected profits (rounded off to 3 decimal places) for the contractors feasible bids are as follows.

| Contractor bid | Expected profit of bid when bidding against contractor XYZ |
|---|---|
| 0.75c | $1.00 \times (-0.25c) = -0.250c$ |
| 0.85c | $0.99 \times (-0.15c) = -0.149c$ |
| 0.95c | $0.96 \times (-0.05c) = -0.048c$ |
| 1.05c | $0.86 \times (+0.05c) = +0.043c$ |
| 1.15c | $0.69 \times (+0.15c) = +0.104c$ |
| 1.25c | $0.39 \times (+0.25c) = +0.098c$ |
| 1.35c | $0.13 \times (+0.35c) = +0.046c$ |
| 1.45c | $0.03 \times (+0.45c) = +0.014c$ |
| 1.55c | $0.00 \times (+0.55c) = +0.000c$ |

It is observed that the bid of 1.15 multiplied by the estimated cost of the project yields the maximum expected profit of $0.104c$. This implies that when bidding against contractor XYZ, over a period of time it would be most profitable for the contractor to submit a bid with a bid-to-estimated-cost ratio of 1.15. For example, if the estimated project cost was $100,000, the bid proposal should be $115,000. Considering the possibility of not winning the contract, the contractor's expected profit for such a bid would be $10,400. Of course, the contractor should keep his bidding information about contractor XYZ current. The best bid ratio for the contractor to use in a future project against contractor XYZ may change, depending upon contractor XYZ's future bidding performances.

If a contractor was bidding against several known competitors rather than only contractor XYZ, he could formulate his bidding strategy in a similar manner. Let us assume that a contractor knows he will be bidding against two known competitors on an upcoming job, contractor XYZ and contractor UVW. Assume the contractor's information about contractor XYZ is the same as in the previous example. Let us assume that the contractor has also gathered information about contractor UVW's bidding performances, and has calculated the probability of his bids being lower than contractor UVW's. This information along with probabilities of winning versus contractor XYZ are as follows.

| Contractor's bid/ estimated job cost | Probability of contractor's bid winning versus XYZ | Probability of contractor's bid winning versus UVW |
|---|---|---|
| 0.75 | 1.00 | 1.00 |
| 0.85 | 0.99 | 1.00 |
| 0.95 | 0.96 | 0.98 |
| 1.05 | 0.86 | 0.80 |
| 1.15 | 0.69 | 0.70 |
| 1.25 | 0.39 | 0.60 |
| 1.35 | 0.13 | 0.27 |
| 1.45 | 0.03 | 0.09 |
| 1.55 | 0.00 | 0.00 |

To calculate the expected profit of the feasible bids, the contractor must determine the probability that his bid is lower than both contractor XYZ's and contractor UVW's bids. Both of these events are independent. The probability of being lower than XYZ is independent of the probability of being lower than UVW. From probability theory, one may show that the probability of the occurrence of joint events, which are independent, is given by the product of their respective probabilities. For example, the probability that the contractor's bid of $1.15c$ wins (is lower than XYZ's and UVW's bids) is the product of 0.69 and 0.70, or 0.483. Having found the bid's probability of winning the contract, its expected profit is calculated as 0.483 multiplied by its immediate profit of $0.15c$, resulting in an expected profit of $0.07245c$. The expected profits (rounded off to 3 decimal places) for all the bids are calculated as follows.

| Contractor's bid | Expected profit |
|---|---|
| $0.75c$ | $(1.00) \times (1.00) \times (-0.25c) = -0.250c$ |
| $0.85c$ | $(0.99) \times (1.00) \times (-0.15c) = -0.149c$ |
| $0.95c$ | $(0.96) \times (0.98) \times (-0.05c) = -0.047c$ |
| $1.05c$ | $(0.86) \times (0.80) \times (+0.05c) = +0.034c$ |
| $1.15c$ | $(0.69) \times (0.70) \times (+0.15c) = +0.072c$ |
| $1.25c$ | $(0.39) \times (0.60) \times (+0.25c) = +0.059c$ |
| $1.35c$ | $(0.13) \times (0.27) \times (+0.35c) = +0.012c$ |
| $1.45c$ | $(0.03) \times (0.09) \times (+0.45c) = +0.001c$ |
| $1.55c$ | $(0.00) \times (0.00) \times (+0.55c) = +0.000c$ |

Note that the contractor should submit a bid which has a ratio of bid cost to estimated project cost of 1.15. Thus, the contractor should make the same bid he should have made when bidding against only contractor XYZ. However, the expected profit of $0.072c$, when bidding against the two contractors, is less than the expected profit of $0.104c$ when bidding against the single contractor. This is because of the added competition. The more competition a contractor has, the less likely he is to receive the contract. The problem of more than two known competitors is handled in a similar manner. We would not conclude that the optimal bid remains unchanged with increasing competition. In general, the optimal bid will have a tendency to decrease with an increasing number of competitors.

The bidding strategy model can be extended to more complicated applications. While such extensions are not discussed in detail here, other applications include the handling of an unknown number of competitors and the including of uncertainty in the cost of the project.

The bidding strategy model implies a competitive bidding construction contract letting procedure. A growing number of construction contracts are being awarded to construction firms through a negotiation process. The potential construction project owner approaches one or more construction firms and negotiates a contract price for which the construction firm will build the project. Although price may not be the only factor in the owner's selection of a firm (i.e., quality of work, financial reputation, and ability to perform within a budgeted time constraint may be other factors), it is difficult to find an owner who isn't operating within a dollar budget. As such, pricing and bidding strategy models are relevant to the negotiated contract process as well as the competitive bidding process.

In reality, pricing procedures and policies differ little for the construction firm participating in the competitive bidding process or the negotiated contract process. It is in the promoting aspect of marketing that distinct differences exist as to the two means of securing construction work. These differences are emphasized in the following section.

## 3.7   Promoting the Firm and its Work

The term promoting is more commonly referred to as selling. In fact some definitions of marketing limit themselves to selling or promoting. Before the firm can "sell" its product, it must first plan its product and price its product. As to the contractor firm, the firm must first determine what type of work it is to construct and where it is to operate. In addition, it must cost or price its services. However, the marketing process is only half complete upon pricing. The contracting firm must now sell its plan and price to a potential client if it is to complete its operational cycle.

The ability to find work through promotion of the firm and its product is the key to free enterprise. Unfortunately the technical skills of the owners of a firm are wasted unless the selling or promoting function is accomplished.

Numerous books have been written and theories developed as to how to "sell". The fact remains that selling can only in part be learned. Personalities, motivation, and verbal skill, while difficult to learn, are vital to selling one's product.

The selling of one's product start with the searching out of potential clients. The means by which the construction firm becomes aware of construction projects are referred to as bid searching. Included in these means are the following.

1. Membership in technical associations
2. Subscription to services
3. Attention to public announcements
4. Personal contacts
5. Advertisements
6. Direct mail
7. Publicity

As is true in most professions, several construction related associations are readily available for the construction firm to become a member. National associations such as the Associated General Contractors (A.G.C.) are available as well as local city and county associations. While the objectives of the many construction associations are numerous and somewhat vary from one association to the next, many serve as a "clearing-house" for the disclosure of upcoming construction projects to their membership. This service in itself can in many cases justify the membership dues that the construction firm pays to the association.

Subscription to construction related services can also aid the construction firm in the finding of projects. New internet services are becoming available such as that offered by the construction associations.

In many cases the least cost and most effective means of searching out work are through attention to announcements and through personal contacts. Municipal, state, and federal construction projects are openly advertised through newspapers, and bulletins. State boards such as the Illinois Capital Development Board notify registered contractors within their state of all upcoming construction within the jurisdiction of the board.

As is true in obtaining work in any business (i.e., especially in service oriented businesses), personal contact is in many cases the key to finding construction projects. Contracts through clubs, church organizations, and charitable organizations can lead to a contract for work. The key to securing work through personal contact is the follow-up. Many times a contractor firm has lost a potential project because of its failure to follow up a lead. In the absence of the contractor firm following up its initial contact, the owner may be approached by yet another firm or the owner may be influenced by others in its evaluation of the feasibility of the project.

Unlike a few service oriented businesses, the "code of ethics" of the construction industry does not prevent the construction firm from openly advertising for work. Yellow page telephone directory advertisements are the most common form of contractor firm advertisements. However several firms place a regularly appearing advertisement in a local newspaper, church related publication, or local news magazine. Others advertise in more widely circulated technical magazines such as *Engineering News Record*, or a country wide newspaper such as the *Wall Street Journal*.

Many a contractor firm has seen fit to sponsor local sport teams such as a little league baseball team or a bowling team. While this practice may stem merely from the firm's interest in the sport or the participants, in many cases the motivation for the sponsorship is advertising. While it is difficult to measure the dollar benefit associated with having the name of the sponsoring team on the uniforms of the team, surely such publicity contributes to the marketing of the firm. The benefit of the type and amount of advertising varies depending on the type of work the construction firm performs.

Similar to the use of advertisements for finding clients, is the use of direct mail. In fact, the use of direct mail can be viewed as a form of advertising. The only difference is that direct mail represents a concentrated effort aimed at known individuals or firms whereas the type of advertising discussed previously was aimed at the general public. Obviously direct mail proves advantageous when in fact the recipient is contemplating the building of a construction project. The fact of the matter is that seldom will an individual or firm initiate a construction project due to the direct mail of a construction firm. There are exceptions as in the case of a homeowner being "sold" on the need to make some home repairs or remodeling through an advertisement or direct mail of a construction firm. Nonetheless, direct mail can prove an economic means of having one's firm added to the list of firms considered when an owner's future construction plans become reality.

Other than personal contacts, direct mail provides the firm with the most personal means of communication with future clients. However the direct mail communication has to be well worded and in fact be aimed at the needs of the client. A poorly written form letter may not only prove non-personal but it will likely be discarded as another piece of time consuming mail.

There is a degree of risk associated with direct mail. Its very existence may lead the recipient to believe the sending firm is hard pressed for work due to unfavorable characteristics of the firm. The end result is that direct mail advertising has to be well written and well directed if it is to result in benefits.

A firm's success has a way of generating good publicity. If this is rec-ognized, and the publicity is properly utilized, the publicity can become part of the firm's marketing program to ensure future success. As to the construction firm, good publicity can develop from several facades of the firm's operations. The building of a unique structure, or the use of new construction techniques can aid the firm through favorable publicity. Other recognized favorable activities include completing a project well before the forecasted completion date (i.e., especially noteworthy when it is a project that has public attention such as a recreational facility or highway), the receiving of professional awards as to individual firm members or the entire firm, the election of members of the firm as officers in professional organi-zations, and the firm's contribution to community or charitable projects.

The firm should not go out of its way to prevent favorable publicity. Regardless of how humble the individual firm members are, publicity should be as much as a part of a construction firm's marketing program as any bird-dogging practice. Seldom if ever will good publicity hinder the firm's marketing. And best of all it is free. Free in the sense that its distribution is free. Naturally it is not free in the effort generated that in fact creates the publicity.

Publicity outlets may be through the local press, trade publications, and professional magazines. While it is gratifying to have these outlets come to the firm there is nothing wrong with the firm initiating the com-

munication to the publicity outlet. The preparation of news releases is the
responsibility of the marketing function within the firm. An effective communication link with the local press can prove to be an inexpensive means
of obtaining construction work through publicity.

Just as good publicity can aid in the finding and securing of construction work, bad publicity can hinder such efforts. Years of sound business
practices, and success in building projects to specifications and within an
allotted time schedule can be forgotten by the public through a single news
release citing an unfavorable activity of the construction firm. An unfavorable suit stemming from the injury or death of an employee, failure to meet
project specifications or a project completion time, and the involvement of
company personnel in bid fixing or bribery efforts can virtually ruin a firm
as to its marketing efforts. In fact is fair to say that a single bad publicity
news release has a much more unfavorable impact on the firm's marketing
than the favorable impact of a good publicity news release. As is true of
many areas of reader interest, bad news seems to get more attention than
good news. The fact that a contractor firm completes construction of a
project before the forecasted completion time often goes unnoticed. However, the public is up in arms should the project be delayed. Such is the
difficulty of obtaining and maintaining marketing benefits from publicity.

Needless to say the contractor cannot spend unlimited time and money
on each of the means noted for finding construction work. Too much effort
and cost expended defeat the very purpose of doing business. The end result
is that the firm must weigh the expected benefits versus the cost for each
and every means considered.

The promoting or selling aspect of marketing does not terminate with
the finding of potential construction project owners. The promoting aspect
of marketing is only complete upon the forming of a contract by the contractor with a client in which the firm agrees to build the client's project
in return for monetary rewards for its services.

In the competitive bidding process, once the potential client is found,
the rest of the promoting process is straightforward. Assuming the contractor meets the qualifications stated by the client, the construction firm
submits its bid which is evaluated as to dollar value by the owner. Normally
the low bid is selected as the bid awarded the contract. The end result is
that once the firm finds the potential client, and satisfies the client's stated
qualifications, its success in the competitive bidding process is dependent
solely on its pricing practices.

The success of the contractor firm in securing a project contract in a
negotiated contract environment is typically dependent on an extended list
of owner considerations. That is, the project owner may recognize factors in
addition to the cost for which the construction firm will build the owner's
project. Included in these factors are the following.

1. Prior dealings with the construction firm.
2. Reputation of the construction firm as to ability to perform the type of work being considered.
3. Reputation of the construction firm as to ability to perform work within budgeted time.
4. Social acquaintances of owner with members of the construction firm.
5. Hiring practices of the construction firm (e.g., affirmative action program of the firm).
6. Recognition of "home location" of the firm (i.e., an owner may choose to deal with a local firm).
7. Recognition of the current work load of the construction firm.
8. The construction firm's willingness to "give and take" with owner as to current and future work negotiations.

Other factors might be present in isolated cases. The point to be made is that the negotiated contract process more clearly pinpoints the need for marketing considerations by the construction firm. Neither the construction firm or the owner is locked in to a predetermined structured bidding process. The contracting firm's external relations are fundamental to securing work in the negotiated contract awarding process.

## 3.8 Placing the Product

The marketing process is completed with the placing of the firm's product. As to the firm, the placing of its product entails the completion of the building process in a manner consistent with the plan and price previously determined. In this sense the placing aspect of marketing is part of a complete cycle starting with planning, continuing with pricing and promoting, and finishing with placing.

The placing aspect of marketing is aimed at satisfying two related objectives. First of all the placing or building of the construction project has to be carried out according to plan and price such that the firm realizes a profit. This profit is vital to the overall objective of a profit oriented firm. Secondly, the placing or building of the construction project has to proceed in a manner that meets the satisfaction of the owner. As discussed in the previous section, the ability to carry out the future promoting aspect of the marketing of the firm is dependent on the reputation of the firm as to past performances. Satisfied customers are the best marketing asset available to the firm. Goodwill is often priceless.

In a number of cases it may prove wise to weigh the placing objective of satisfying the customer above that of profit. A construction firm's willingness to absorb added costs associated with an owner's change order may result in the securing of future contracts with the owner. Naturally the construction firm cannot always give and not take. Profit remains its vehicle for successful operations. However a short-term loss that creates owner satisfaction may in fact lead to long-term profits. Each project and individual circumstance is unique as to each give and take decision. The important point is to recognize the long-term as well as short-term effects of each project and individual circumstance.

Placing of the construction project consistent with the predetermined plan and price entails the utilization of sound project management tools and practices. Economic material procurement and project material flow, effective equipment utilization, productive use of labor, recognition of weather as to resource scheduling, the training and utilization of management personnel, and the ability to coordinate with project subcontractors are all necessary to placing the construction project.

Planning, pricing and promoting provide the contracting firm the potential for the carrying out of its profit oriented objective. It is the placing aspect of marketing that provides the firm the realization of such an objective. To this end, the firm's marketing and overall organization structure should fully recognize the importance and interrelationship of placing relative to the more commonly though of aspects of marketing.

## 3.9   Summary

Of all the business functions that are necessary to carrying out a profitable firm, marketing is perhaps the most difficult to learn and quantify. Marketing is a people problem, a communication problem. The firm with all the technical and management know-how in the world will not generate profits unless it can market is product or service.

The complexity of marketing as to the contracting firm is increased by the fact that the construction industry is somewhat unique as to its type of business and product. The construction firm is somewhere between being a manufacturer of a product and being a service oriented business. The firm builds a product, but is not truly its product in that it is constructing a predetermined design of the owner. Typically the construction firm will build a given type of product only once in that each construction project is somewhat unique. The end result of the uniqueness of the construction industry's product is that many of the marketing theories and practices that apply to other industries are incompatible with construction firm marketing.

The marketing function is more than that of selling. In particular, marketing consists of planning, pricing, promoting (more commonly referred to as selling), and placing. These four aspects of marketing make up a complete cycle. Absence of any of the four results in an ineffective marketing program.

Planning consists of determining the type of work, the location, and the amount of work the firm would like to perform. Pricing quantifies the profit oriented objective of the firm such that the firm is more than doing work for the sake of doing work.

Promoting or the selling aspect of marketing provides the link between plans and reality. The contracting firm must "sell" its plan and price. Finding work is of utmost importance. Numerous means such as personal contacts and advertising are available to the firm for the purpose of finding work. Once an owner is found, promoting efforts differ depending on the contract awarding procedure. Low cost is the key in the competitive bid process. While low cost remains important, external relations increase in importance as to securing a contract with an owner in a negotiated contract environment.

Once a project contract is secured, the marketing function should see to it that the project is placed consistent with the predetermined plan and price. This aspect of marketing focuses on project management. Planning, pricing, and promoting provide profit potential. Placing realizes it.

# Chapter 4

# Financing

## 4.1  Introduction

Financing for projects and operations is one of the contractor's greatest concerns. Historically, contractors operate on a small profit margin with little working capital to carry them from job to job. This is the case because of retainages carried through a job that may last several years, large payrolls that must be met every week, and receipts that generally occur only once a month. Most contractor's find, because of these factors, that it is necessary to obtain some sort of short-term financing from time to time. The tasks of budgeting, cash flow, and determining financing are needed by the contractor.

In addition to the need for cash for financing of project costs, the contractor may need cash for the financing of equipment purchases and for financing planned company growth. Cash for these purposes is usually required for a longer period of time than cash required for project cash flow.

Funding requirements are often classified as to the period of time for which the funds are obtained. Thus, funding is often classified as to short-term, intermediate-term, or long-term. There really is no clear-cut line between what is a short-term funding and what is an intermediate-term funding or as to what is an intermediate-term funding and a long-term funding. However, for purposes of discussion it is useful to differentiate funding as to short-, intermediate-, or long-term. For our purposes we will consider short-term financing to be of less than one year duration, intermediate financing to be of duration from one to ten years, and long-term financing to be of duration greater than ten years.

## 4.2 Short-Term Financing

Short-term financing is usually identified with a contractor's need for funds for the payment of costs associated with the building of a construction project. Typically the contractor firm receives payment for its work from the project owner after the construction firm has incurred the costs of doing the work.

The payment to the contractor firm for projects are based on regularly occurring progress reports and an owner retainage clause. For example a contract may read that the contractor firm is to be paid at the end of every month for the work judged to be completed at the end of the previous month. However the owner is to retain 10% of each payment until the project is completed. This "holding back" of a part of the payment is intended to ensure that the project is completed according to specifications

Regardless of whether the contractor firm is involved in building a residential unit or performing commercial, industrial, and heavy and highway construction, the firm typically receives its revenue after the incurrence of a cost liability. This liability may in fact have to be paid before the revenue is received. The end result is that the firm has a need for short-term financing.

## 4.3 Trade Credit

Probably the major source of short-term financing available to the contractor is the credit extended by material suppliers. This form of financing is referred to as trade credit. Although it is difficult to document the actual usage of trade credit it would indeed be a rare event to find a contractor who did not take advantage of some sort of trade credit.

Trade credit is the credit extended by a seller to a buyer of goods that the buyer will ultimately resell. This definition excludes the credit given to a contractor for the purchase of equipment, in that the intent of the contractor is to use the equipment, not resell it. In addition, trade credit does not include common credit in that this form of credit relates to goods for consumption rather than resale.

Trade credit is especially relevant to the contractor in that the contractor is a major purchaser of material when building a construction project. In addition, the relatively small size of the average contractor is related to the wide usage of trade credit in the construction industry. It has been shown that smaller firms make a greater usage of trade credit than larger firms. Whereas large firms usually only borrow under favorable conditions, small firms including the average contractor, often have to rely on trade credit when short-term money is too expensive to obtain.

Trade credit differs from other means of financing, such as bank loans, in that one may not have to pay added costs associated with its use. If all available cash discounts are taken, the use of trade credit adds no cost. For this reason, it is desirable to take advantage of this form of credit. One pays something extra only upon failure to take advantage of the discount.

- The arrangement between the purchaser of material and the supplier does not always allow for a cash discount. For example, the trade terms may be Cash Before Delivery (C.B.D.) or Cash On Deliver (C.O.D.).
- A more common trade arrangement between a contractor and a supplier is referred to as ordinary terms. Such an arrangement provides for a cash discount if the bill is paid within a stated number of days, such as 10 days. The payment must be paid before another stipulated period, such as 30 days, or else a penalty interest charge is imposed. Such an arrangement, assuming a 2% cash discount, is designated as 2/10, n/30. If the material is shipped from a distant location or the shipment is slow, the terms may be 2/10, n/30 A.O.G. (arrival of goods).

Another somewhat frequently used form of trade arrangement is the use of monthly billings. This arrangement is especially advantageous when it is difficult or uneconomical to keep track of cash discounts on each transaction. The monthly billing arrangement allows the purchaser to take advantage of a cash discount for all his purchases in a given month, if the payment is made by a given date of the following month. For example, the terms 2/10, E.O.M. n/30 provides for a 2% discount on all material purchases in April, if the bill is paid by the 10th of May. In addition, the entire bill must be paid by the end of May or a penalty interest charge is imposed.

## 4.4  Trade Credit and the Contractor

The relationship between the contractor and the material supplier should be one of trust. This is because each depends on the other for their making of a profit; therefore, it cannot be one of untrustworthy actions. There has to be a great deal of mutual understanding with much give and take on both sides.

The financial arrangements between a contractor and a material supplier are very much standardized throughout the industry. The more common arrangements consist of ordinary credit and monthly billings. Generally as far as actual numbers are concerned, most suppliers give a 2% discount for cash payment. It should be noted here that this 2% is from the quoted price and not from the list price of the material in question. A word later on the difference of the list price to the quoted price. On any unpaid debt left on the books after fifteen days and paid before the thirtieth day there is no difference from the quoted price. This again is used as an inducement to pay the debt early for after thirty days, a 1.5% service charge per month is levied against the unpaid balance. This service charge is not used for the cost of carrying the debt on the books for there is hardly any expense in that, but is used in consideration of the time value of the money. It works something like this: After a large order has been shipped, the supplier must replenish any depleted stock. This takes money for materials, labor, and overhead. To get this money the supplier may be forced to borrow money which must be paid back with interest. Another alternative may be for him to use his own capital which he must remove from some other area which is providing him with some opportunity interest which he now forfeits. To make either of the above methods economically practical a service charge must be raised to provide some security and financial gain to the supplier for either risking his own money and forfeiting his interest gain or having to borrow money at some interest rate.

There are exceptions to every rule and the supplying business is no exception. If a contractor has dealt regularly for some time with a certain supplier and paid his debts on time, the supplier, relying on previous experience, may take over some of the risk of the debt and waive the 1.5% charge in an expression of good faith while also hoping to keep the customer for many years to come. The supplier rationalizes this action by stating that the loss of the 1.5% is more than compensated for by having this person become or stay a repeat customer.

As can be seen, material suppliers do extend some free credit, varying from ten to thirty days. The contractor is wise to take advantage of this free credit as much as possible to keep his working capital at a maximum.

As was previously mentioned, there is a difference between the quoted and list prices of a supplier's materials. A trade discount is used for many varied reasons under many different conditions. This discount ranges anywhere from 5% to 40% off the list price. Therefore the price given with the discount is the seller's quoted price. The primary reason for the discount is that among suppliers of the same material the list prices are almost exactly the same and so to compete for business the discount method is used. The amount of the discount varies with the time a contractor has been with a particular supplier; with the highest discounts being given to the oldest customers with good financial backing. The important aspect of the mutual understanding relationship is that these discounts can often spell the difference between a profitable or unprofitable construction project.

## 4.5   The Cost of Trade Credit

The failure to take trade discounts results in an interest rate associated with the use of trade credit. The effective interest rate varies depending on the terms of the supplier, and the time at which the payment is made. For example, let us assume a contractor purchases $1,000 of materials from a supplier on terms of 2/10, n/30. If the contractor pays the bill on or before the 10th day he only pays $980, thus, saving $20. However, if paid after the 10th day the contractor has to pay the extra $20. Let us assume that the contractor pays the $1,000 on the 30th day (thus avoiding a further penalty cost). In effect the contractor is using the supplier's money for 30 days. The effective interest rate for the last 20 days is 20/980, and there are 365/20 20-day intervals during the year. The true annual interest rate of missing the cash discount and paying the bill on the 30th day is:

$$(20/980) \times (365/20) = 0.3724 \quad \text{or} \quad 37.24\%$$

If the contractor were to miss the discount (pay after the 10th day) but paid before the 30th day his effective interest rate would in fact be more than the rate previously computed. This should be obvious in that the contractor is paying the $1,000 prior to when it is due. Thus, the contractor is losing an opportunity to use the money, which results in a higher interest rate. The effective interest rate for payment on any day after the discount day and before the penalty date can be calculated by multiplying the effective interest rate for the total period, 20/980, by the ratio determined by 365 days divided by the number of days beyond the discount date. For example, let us assume that the contractor paid the $1,000 bill on the 20th day, 10 days after the discount date. The effective interest rate associated with not taking advantage of the discount is calculated as follows:

$$(20/980) \times (365/10) = 0.7448 \quad \text{or} \quad 74.48\%$$

Note that this effective interest rate is double the rate associated with paying the bill on the 30th day.

It should also be noted that if the contractor pays a bill before the discount date, he is in effect paying an added cost in that he is providing his cash to the supplier before it is due. Thus, the contractor should either pay his supplier on the discount date or on the day before a penalty charge is imposed. Of course the discount date is preferable.

## 4.6   Short-Term Loans

Next to trade credit, short-term bank loans are the most widely used short-term financing for a contractor firm. Short-term bank loans are often unsecured. However, the borrower may have to pledge or sell some of his assets as security, in which case the loan is secured. Commercial banks are the major source of secured and unsecured short-term loans. Finance companies are another source of these loans. Finance companies may be of several types. A consumer finance company usually only gives small loans to individuals. Commercial finance companies give loans to a company in return for the company's pledge of accounts receivable or inventory. Factors are finance companies that deal only in granting loans for accounts receivable security.

The security that banks or finance companies require companies borrowing money to pledge may be cash, inventory, or accounts receivable. Banks may require a borrower to provide cash security by means of having the right to offset against any deposits the borrower has in the bank. Let us assume that after having obtained a $10,000 short-term loan from a bank, a contracting firm has the following balance sheet.

| Cash in bank | $5,000 | Bank Loan | $10,000 |
|---|---|---|---|
| Fixed assets | 35,000 | Accounts Payable | 10,000 |
| | | Capital | 20,000 |
| | $40,000 | | $40,000 |

If the contractor is forced to liquidate and the fixed assets bring $5,000, the bank's claim would be $5,000, calculated as follows.

| Available | | Distribution | |
|---|---|---|---|
| Cash available | $5,000 | Bank loan | $5,000 |
| Cash from liquidation | 5,000 | Accounts payable | $5,000 |
| Available | 10,000 | Distribute | 10,000 |
| | $10,000/20,000 = 0.5$ | | |

However, if the bank had the right to offset, upon contractor liquidation the bank would receive $6,667, calculated as follows.

| Available | | Distribution | |
|---|---|---|---|
| Cash available | $5,000 | Bank loan | $1,667 |
| Cash from liquidation | $5,000 | | 5,000 |
| | | Accounts payable | $3,333 |
| Available | $10,000 | Distribute | $10,000 |
| | $5,000/15,000 = 0.333$ | | |

Thus, upon the contractor's inability to repay the loan, the bank receives $1,667 more with the right to offset versus the situation where such a right does not exist.

When inventory is required as the security for a short-term loan, the amount of money loaned is often a percentage of the inventory. For example, a lender may have a policy of lending up to 70% of their inventory value of the borrower. The inventory that is used as security may be kept in the possession of the borrower, or it may be held by a third party. The difficulty associated with an agreement whereby the borrower keeps possession occurs when the borrower manipulates the inventory in a manner that reduces the required security inventory. Such a situation places the lender in a weak position. For this reason the lender may require the borrower to place the inventory in the possession of a third party where it is controllable and under the supervision of the lender.

A borrower may also provide security for a loan by assigning accounts receivable to the lender. The lender, as with the inventory, will usually only loan a percentage of the value of the assigned accounts receivable. This percentage is usually higher for commercial finance companies than it is for commercial banks. For example, a typical commercial finance company may loan 80% of the accounts receivable, whereas a typical commercial bank would more likely loan about 75%.

Other forms of collateral used for security include securities, mortgages on the borrower's residential property, and passbooks on savings accounts. Stocks and bonds are other sources of collateral. To be used as security, they must be marketable. Banks will usually borrow no more than 75% of the market value of quality stocks, and 90% on Federal municipal bonds. Life insurance may also be an acceptable form of collateral. Banks normally will lend up to the cash value of such a policy.

Because of their weak financial condition, a borrower may get another individual to sign a loan note to strengthen the borrower's credit. The individual who signs such a note is referred to as an endorser, and he becomes contingently liable for the note. An endorser may be a co-maker of a note, or he may be a guarantor. A co-maker is an individual who creates an obligation jointly with a borrower. He becomes equally liable with the borrower for the loan. A guarantor is one who guarantees payment upon default of the borrower.

In addition to requiring collateral, a bank upon making a loan to a borrower, may place restrictions on certain management practices of the borrower. For example, the lender may limit the dividends that a borrower can pay out to stockholders while the loan is outstanding.

As previously noted, short-term loans are referred to as short-term financing in that the loans are obtained for a period of a year or less. Yet another description of short-term loans is that they are repaid within the normal course of business. For most industries these descriptions are consistent in that the normal business cycle is usually one year or less. However, in the case of a building project that takes several years to construct, the business cycle (the project duration) will be a period greater than one year. In such a case, the financing for the project is referred to as short-term financing.

A short-term loan from a bank may be in the form of a single loan or it may be a result of an arrangement referred to as a credit line. A credit line is an informal agreement between a bank and borrower as to the maximum amount the bank will lend the borrower at any point in time. The agreement is not legally binding on the bank. The bank often imposes a fee for a firm commitment to keep a credit line for a borrower.

A credit line is usually reviewed annually by the bank. In addition, many commercial banks require borrowers to keep some percentage of the credit line as a deposit in the bank. This deposit is referred to as a compensating balance. The compensating balance requirement has the effect of increasing the borrower's effective interest rate associated with his bank loans. Finally, several commercial banks require a borrower to annually reduce his debt to zero in order to maintain a credit line for the next year. This annual elimination of debts is referred to as cleaning up the credit line.

## 4.7   Bank Loans and the Contractor

The first and foremost avenue for the contractor to consider for a short-term loan is his local bank. Initially, a contractor is wise to do business through his various accounts with a large bank. The bank and the contractor need to have a mutual trust and familiarity for each other. A contractor will do well to have his business all in one bank and to build a good reputation with his particular bank.

Short-term contractor financing through a bank may be secured with collateral or unsecured. Banks are extremely careful about lending money to contractors. One reason is that contractors, in general, have been found to be bad risks. Because of the many variables involved with a large project, it is difficult to determine whether a loan may be considered safe by the bank. Another reason for the reluctance of banks to make loans on large projects is because of the uniqueness of the product.

When a contractor goes to a bank or commercial lending institution for a loan, the bank will base its analysis on several factors. Factors that the bank may consider are: risk involved with the particular project; reputation of the contractor; type of job and who the owner is; other types of similar work going on in the area; examination of documents, such as take-off sheets, CPM, profit percentage; current financial statements of the contractor; the workload the contractor is presently engaged in; and the ability of the contractor to perform the work involved. Before extending any type of financing to a contractor, a bank will do an analysis involving the above-mentioned factors. The process of financing should not be viewed as trying to "trick" the institution into lending the money. The contractor should present his needs and reasoning, as well as showing the safeness of the loan to commercial lending departments in a professional manner. Financing should be viewed as an exchange of information and resources: the contractor showing potential and the banker showing his available dollars. A bank will base its decision on fairly reliable concrete facts. A large lending institution will be involved in what is going on in the area and will probably have more facts available to them than the contractor in their financial decision analysis. A refusal for a loan for a particular project may, in the long run, save the contractor problems, as well as money.

In evaluating the feasibility of giving a loan to a contractor, a banker may also evaluate the firm by means of financial or operating ratios. A working capital ratio of 2 to 1 and a quick ratio of 1.25 to 1 are usually considered acceptable. Ideally, the contractor should have on hand working capital that is 10% to 20% of the value of his ongoing projects.

The most common form of short-term financing from a bank is the 90-day loan. A bank may be willing to loan up to the net worth of a company. The effective rate would include the interest rate plus direct costs involved with the loan. Construction loans do carry a higher interest rate than other commercial loans. A contractor is wise to take advantage of the short-term financing from material suppliers and then to borrow from insurance policies before going to a bank.

## 4.8   Cost of Short-Term Loans

The interest rate charged by banks for short-term loans is very dependent on the prime rate of interest and the availability of money. The prime rate is the interest rate banks pay for money. When money is tight, the spread between the prime rate and the interest rate charged to borrowers from banks increases.

The interest rate on short-term loans is also dependent on the size of the loan. The smaller the loan, the higher the rate; the reason for the higher interest rate for small loans is related to the fact that the bank has certain fixed costs associated with making a loan. For a small loan, these fixed costs make up a larger percentage of the loan, resulting in the banks charging a higher interest rate.

The effective interest rate actually paid by a borrower is dependent on the loan agreement. The simplest form of agreement is one where the borrower pays the lender an interest rate on the unpaid balance of the loan. Thus, a borrower upon making payment on a loan one year after receiving $5,000 at 8% a year would make a payment of $5,400. The 8% interest rate is in fact the effective interest rate.

A more common loan agreement used by banks is the discounted loan. The borrower signs a note promising the bank to pay $5,000 to the bank one year from now. The bank, assuming an interest rate of 8%, takes out the interest charge in advance and thus gives the borrower $4,600. The effective interest rate for this discounted loan is $400/$4,600, or approximately 8.7%.

Yet another common loan agreement is the installment loan. This type of loan is often combined with a discount agreement. For example, a borrower upon borrowing $5,000 at 8% a year would receive $4,600.

The bank would then require the borrower to repay back the $5,000 in equal twelve month payments. Thus, at the end of each month the borrower would pay the bank $416.67. The effective interest rate for such a loan is considerably higher than the 8%. For one the loan is discounted. Secondly, the borrower does not have the use of the $4,600 for the entire year in that he repays it in twelve installments. An approximate effective interest rate for the loan can be calculated by dividing the interest charge by the average amount of money the borrower has from the loan. Thus, the approximate effective interest rate for the loan in question is $400/$2,300, or approximately 17.4%. A more accurate method of calculating the effective interest rate is given by the following formula.

$$i = \frac{(2m\,D)}{P(n+1)}$$

Where $i$ is the effective annual interest rate, $m$ is the number of payments in one year, $D$ is the interest charge in dollars, $P$ is the cash advance, and n is the total number of payments. For the example in question the effective annual interest rate is calculated as follows.

$$i = \frac{2 \times 12 \times 400}{4,600 \times (12+1)} = 16.1\%$$

Regardless of whether the approximate or theoretically correct method is used, it can be seen that the effective interest rate is much greater than the 8%.

# 4.9   Other Sources of Short-Term Funds

In addition to trade credit, and bank loans, other sources of short-term funds available to the business firm include the sale of the firm's accounts receivable; loans from relatives, friends, customers, and suppliers; and loans from the Small Business Administration (SBA). These methods of obtaining funds may be advantageous to the contractor.

Several firms, rather than providing their accounts receivable as security, actually sell their accounts receivables to companies called factor companies who purchase the accounts at a discount. The purchasing company then has the ownership and responsibility of collecting the accounts receivable. Obviously due to the effort and risk associated with making the collections, the factor charges a rather high interest and service charge for these services.

The contractor is usually not active in the selling of his accounts receivable. This practice is more widely used in an industry such as the textile industry which has wide seasonal variations in volume and accounts receivable.

Another form of short-term financing that is available to the small contractor is the loan from the SBA. The SBA was created by Congress in 1953. It was established to aid and stimulate small business. In regard to

the construction industry, a contractor is classified as a small business if
the annual receipts of the firm are not more than $5 million, averaged over
a three year period. Because of the small size of most contractors, many of
the existing firms are eligible for aid from the SBA. This aid often comes
in the form of loans for working capital (short-term loan).

Usually loans from the SBA are only available when other sources have
been shown to be unavailable. The SBA is involved in giving out two types
of loans. For one a borrower may be able to obtain a direct loan from the
SBA. This type of loan is obtained directly from the SBA. A second and
more frequent type of loan which a firm can obtain from the SBA is a
participation loan. In participation loan the SBA shares the loan with a
bank or guarantees the portion of the loan provided by the bank. Thus,
the small firm that is unable to provide security and thus cannot obtain a
bank loan, by means of a guarantee from the SBA may find that such a
loan becomes accessible.

Because of the contractor's inability to provide security for bank loans,
SBA financing is often used by the construction industry. In addition to
providing working capital loans, the SBA involves itself in loans for business
expansion and for the purchasing of equipment.

## 4.10    Intermediate-Term Financing

Intermediate-term financing was previously described as financing that was
arranged for a period of more than one but less than ten years duration.
Intermediate-term financing is frequently referred to as term financing or
term loans. Because of the longer duration of the financing, intermediate
loans are used for a different purpose than are short-term loans. In particular, intermediate financing is associated with the purchase of machinery
and equipment, and for the permanent increase in the firm's current assets
such as material inventory. Unlike many short-term loans, term loans are
almost always secured by collateral of the borrowers. The required types of
collateral are similar to those discussed for short-term loans.

Term loans are repaid by means of monthly, quarterly, or yearly installment payments. They are amortized in the same manner as home mortgages. That is, the initial payments are almost entirely interest, resulting in
little reduction in principle. Latter payments go more to reducing the loan
principle. When a firm takes out a term loan it expects to repay the loan
from profits generated from the usage of the newly acquired assets. Thus,
the lender has to evaluate the firm's ability to generate these profits when
determining the feasibility of making such a loan. An evaluation of several
consecutive income statements and balance sheets of the firm becomes a
means of evaluating the risk of the loan.

One may question why term loans are chosen rather than short-term loans when purchasing equipment. The most obvious reason is that the firm has to be certain of the availability of the funds. The firm cannot be assured of receiving continual short-term loans from a bank. In addition to the changing financial structure of the firm, the bank's cost of making several loans rather than one, and the bank's uncertainty as to the availability of money would make future short-term loans uncertain. In addition, the term loan often affects a firm's balance sheet in a more favorable manner. A term loan appears as a long-term liability on the balance sheet, thus it doesn't totally influence the firm's current liabilities and this does not affect the firm's current ratio (current assets/current liabilities). A low current ratio often is a constraint to the firm's ability to obtain financing, especially term financing.

## 4.11   Sources of Intermediate Financing

Banks do not play as dominant a role as a source of intermediate financing as they do for short-term financing. Banks favor the giving of short-term loans in that the giving of intermediate loans results in constraints upon the bank's deposit liabilities.

Banks almost always require the term loan to be secured by assets. Because of the duration of the loan, the bank has to make safeguards against the borrower's use of the assets. This is done by placing a constraint upon the borrower which prevents him from selling certain fixed assets. In addition, the borrower is prevented from pledging the assets to another lender for the procurement of a loan.

In addition to constraints placed upon the borrower's assets, the bank also takes steps to ensure itself that the borrower will meet his installment payments on the term loan. This is often done by means of prohibiting the borrower from making withdrawals or paying dividends in excess of the amount that reduces the firm's net working capital or current ratio below a stated minimum.

A bank usually reserves the right to collect the entire loan before it is all due if the borrower fails to make an installment payment that is due. This is done by means of an acceleration clause in the loan agreement.

Although savings and loan associations do not make unsecured short-term loans, they do provide term financing along with long-term financing. Financing by savings and loan associations require real property as collateral. As such, savings and loan associations seldom aid in equipment financing.

A savings and loan is restricted by its national charter as to how much money may be invested in construction. A large savings and loan association has a tremendous amount of assets and is a good source of money for a contractor. However, since real property is required for collateral, loans are somewhat limited to owner-developers or contractors that are in some sort of legal partnership with the project owner.

As mentioned in a previous section of this chapter, the small sized contractor can turn to the Small Business Administration (SBA) for loans. In addition to supplying loans for use as working capital, the SBA is a source of loans for the purchasing of equipment. Thus, the SBA is yet another source of intermediate financing. However, as in the case of SBA working capital loans, equipment loans from the SBA are usually only available when it can be shown that other sources are not available.

Although not as abundant a source of term loans as are commercial banks, intermediate financing is also available from insurance companies. However, insurance companies tend to give few loans to small firms, especially small contractors. In addition, loans from insurance companies tend to be long-term rather than intermediate-term. Sometimes a bank and an insurance company combine their financing in giving a loan. In such a case the bank provides the intermediate-term funds and the insurance company provides the long-term portion.

A unique source for intermediate-term financing, and a source widely used by the construction industry is equipment financing from finance companies. Equipment can be financed through a finance company by means of purchasing the equipment on installment or by leasing the equipment. In addition to differences in the format of the arrangement, the accounting methods for each of the two procedures differ. As such, the impact of purchasing, whether installment or leasing, differs on a firm's financial statements.

The main difference between purchasing equipment on installment from a finance company and leasing the equipment from the finance company lies in who owns legal title to the assets. Under the purchase agreement, the purchaser (the borrower) obtains title of the asset upon payment of all the purchase installments. Under the lease agreement the user of the equipment (the lessee) never obtains title to the asset. However, he is in effect financing the asset in that he is using it for payments of money. A lease agreement may be written such that the lessee is liable for repairs and maintenance costs, or it may be written such that the finance company (the lessor) is liable for such costs.

Another arrangement by which a contractor can obtain funds through a finance company is by means of a leaseback arrangement. In such an arrangement the firm exchanges its owned assets to a finance company for cash. The firm then repurchases the assets by means of installment payments. Thus, in effect, the firm sells title to its assets and repurchases the asset by installment payments. Obviously the finance company charges the firm for providing the financing. The effective interest rate in a leaseback arrangement is usually quite high. However, in absence of available sources of funds to the borrower, such an arrangement becomes feasible.

## 4.12 Intermediate Financing and the Contractor

The contractor's need for intermediate-term financing is centered around his need for equipment financing. Unlike entrepreneurs in many other manufacturing industries, the contractor has little need for financing for a permanent increase in inventory.

Due in part to the nature of the use, and the inability of the contractor to provide security, a common source of a contractor's intermediate-term financing is obtained from an equipment finance company. This financing is obtained by means of purchasing installment agreements or lease agreements. The past ten years has seen a twofold increase in the number of construction equipment finance companies and leasing companies. Financing for equipment from a finance company is available to many contractors, including those that may not be able to obtain financing from another source due to their inability to provide security.

The inability to provide security loses its value to the lender in the case of an equipment finance company in that the finance company legally holds title to the equipment until the contractor has completed total payment for its purchase. If the contractor fails to meet installment payments, the finance company is not left holding a useless and unsellable asset in that they are in the business of selling and leasing such equipment.

Next to equipment financing through a finance company, commercial banks provide the largest source of term financing to the construction contractor. The availability of these loans to a contractor varies depending on the interest rate on money at a given point in time, the bank's available funds, the ability of the contractor to provide security, and the bank's evaluation of the firm's ability to repay the loan.

Unlike a short-term loan where the interest rate in the banking system remains relatively unchanged throughout the period of the loan, the interest rate may change substantially during the duration of a term loan. As a result, a bank may be unwilling to commit funds at a low interest rate to a contractor for the purpose of term financing. If the interest rate increases substantially the bank may have to pay a higher rate for obtaining cash than the rate it is collecting on its loan. The willingness to lend money at a given rate also depends on the bank's available money. If money is readily available from the banking system, the interest rate consideration lessens in value. A low interest rate and sufficient available cash often occur simultaneously in our banking system.

If money is "tight" the bank may charge the contractor money over and above the current interest rate when making the loan agreement. This extra charge is in the form of points which add to the contractor's effective interest rate.

Intermediate-term loans to a contractor from a bank usually require security. As a result these types of loans are more readily available to larger contractors who have more assets which they can put up as security.

The bank's evaluation of a contractor's ability to repay a term loan is centered around the firm's ability to generate profits over a period of years. Thus, in addition to working capital considerations, the strength of the firm's management and the firm's market capabilities become important considerations. The small contractor is often put at a disadvantage when these factors are evaluated.

The SBA is yet another widely used source of financing to the contractor. However, this source is limited to contractors who do an annual volume of work less than $5 million, averaged over a three year period. In addition, the firm must also be able to show that other sources of money are not available.

## 4.13    Cost of Intermediate Financing

A bank will normally charge a higher interest rate for an intermediate-term loan than a short-term loan. This is due mainly to the uncertainty of the interest rate of the banking system as a function of time. It should be noted that when the interest rate is high, the gap between the interest rate of intermediate and short-term loans narrows. The stated interest rate is also dependent on the size of the loan. Normally, the larger the loan the smaller the stated interest rate. In respect to the actual interest rate charged by the bank for term loans, it should be remembered that these loans are normally repaid in installments. Thus, the effective interest rate is much higher (often double) the stated interest rate. The stated interest rate for the term bank loan is closely related to the prime rate. For example, if the prime rate is 7%, the stated interest rate may be 8%. The effective interest rate is much higher, the actual rate depending on the duration of the loan. On large-term loans that extend for a rather long duration, banks may make an arrangement with a borrower to adjust the stated interest rate as a function of changes in the prime rate.

The effective interest rate associated with term financing from finance companies normally exceeds the effective interest rate of secured-term loans from banks. Term loans for equipment are normally for a duration of one to four years. Knowing the duration of the financing, and the stated interest rate, the effective interest rate can be calculated.

Equipment finance companies have to charge an interest rate high enough to cover their expenses associated with the possibility of buyer default and subsequent reselling of the equipment. If the finance company is in the business of leasing equipment, the interest charge or rent must be high enough to allow for replacement cost, maintenance and repair costs, and provide a reasonable profit.

## 4.14   Long-Term Funding

As previously defined, long-term funding is financing that is intended to be used for a period of ten years or more. Long-term funding can be separated into two sources, long-term debt and equity funding. Yet another source of long-term funding is the obtaining of long-term leases on property or equipment.

Because of the risk to the lender long-term debt and equity funding is usually only available to the large, established firm. This type of firm is usually a corporation. As a result, sources of long-term funding are not often available to the contractor in that he tends to be small in size in regards to owned assets. However, as a contractor grows, as many do, long-term debt and equity funding become possible. Although one thinks of going to an outside lender when speaking about financing, a proprietor upon starting a firm by means of investing several hundreds or thousands of dollars for the purchase of assets is in effect providing long-term funds for the firm. This is in fact the purpose for obtaining long-term funding, that is, long-term funds are obtained for the purchase of assets and for the capital structure that will permit the firm to operate over a long period of time, in particular for more than 10 years.

The reasons for obtaining long-term funds versus attempting to obtain continuous short-term or intermediate-term funds are centered around the uncertainty of obtaining several continuous loans, and the large dollar value of funding involved. Upon implementing a plan to increase capacity by means of building new plant facilities, a firm has to be assured of being able to have the funds available to meet the large financial commitment. In addition, it is not likely that they will be able to repay the financing (if it is debt financing) in a period of less than 10 years because of the large dollar value of the financing. Even if the firm were capable of repaying it in a shorter period, say 5 years, it may be better to defer the payments over a longer period in that large payments over the 5 years will have a substantial adverse affect on those year's earnings. It is usually better financial management if the assets are paid for as the revenue is generated from the usage of assets. Thus, long-term funding is often consistent with matching the firm's revenues against expense.

Long-term debt and equity funding are quite different. Long-term debts are paid back by a borrower to a lender in a fashion similar to that of an intermediate-term loan except that the duration is longer and the payment of the principle often is made in a single payment at the end of the duration of the loan. Equity capital is never repaid. It is money obtained by selling a part of the interest in the business. In addition to the differences in maturity, other differences between debt and equity include the claim on assets and income, and the effects on management.

## 4.15   Equity Funds

Proprietorships and partnerships raise equity capital by means of the investment of the individual proprietor or the investments of the individual partners. Corporations raise their equity capital by means of selling shares of stock, which represent certificates of partial ownership to investors. The funds obtained from these investors do not have to be paid back to the investor as they do when the funds are obtained from the purchasers of bonds.

Because of the small size of the contractor, obtaining funds by the selling of stock is not readily available to most of the firms. This is also true of the obtaining funds by the issuance of bonds. The firm, due to its lack of size in regard to its assets and its operations, finds it difficult to locate individuals or firms who are willing to invest their capital in the firm. In addition, the contractor may be unwilling to sell stock as a means of obtaining equity funds in that the contractor is unwilling to sell part ownership of his firm. It is also true that the contractor usually does not have the same demand for long-term financing as he does for shorter duration financing. He usually does not require large amounts of cash, such as those negotiated in long-term loans for the building of a plant. The contractor is more in need of cash for working capital and equipment purchasing.

## 4.16   Types of Stock

Equity ownership by means of holding stock in a firm differs greatly from being a creditor of a firm by means of holding a bond of the firm. The creditors have first claim on the income of the firm, and they have first claim on the assets of the firm upon insolvency. On the other hand, creditors, other than constraints that they may place upon management in the loan agreement, have no voice in the management of the firm. Stockholders have a voice in management in proportion to the percentage of the number of shares of stock they hold. In addition a firm does not have a legal obligation to pay dividends to stockholders as it does have an obligation to pay interest to bondholders.

In addition to differences between debt and equity, there are differences between different types of equity ownership. Securities issued by a corporation may be either preferred stock or common stock. The major difference between the two types of stock is that preferred stockholders have a prior claim to dividends before dividends can be paid to the common stockholders. On the other hand, common stockholders often have a larger share in the voice of management.

Preferred stock may be either cumulative or non-cumulative. If it is cumulative, dividends not paid in a given year must be paid before common stock dividends are paid. Preferred stock may also be participating. In this case after the stated amount of dividends are paid to the preferred stockholders, excess dividends are shared by the preferred and common stockholders.

## 4.17   The Cost of Equity Funds

The calculation for determining the interest rate or cost of capital for the issuance of stocks is quite simple and determinate for preferred stock, but more difficult and uncertain for common stock. The market price at which preferred stock can be sold is usually reliably predicted by comparing the issuance to competing issuances. In addition, unless the firm is not being profitable, it will issue regular yearly dividends of a known amount to the preferred stockholders.

The calculation of the interest rate is not complicated by taxes in that dividends on preferred stock is not a deductible expense. Thus, if a firm receives $50 for the issuance of a share of preferred stock and pays an annual dividend of $3 to the stockholders for each held share, the interest rate or cost of capital to the firm is calculated as 6% ($3/$50).

The calculation of the interest rate or cost of capital for issuing common stock, is complicated by the uncertainty concerning the market value of the stock, and the uncertainty of the frequency and amount of annual dividends paid. The market value at which the common stock will be purchased by investors is affected by the investor's prediction of the firm's future earnings and dividends, and investor speculation concerning the overall value of securities. The uncertainty is made more difficult by the fact that the issuance of new shares dilutes the existing shares in regard to earnings per share of outstanding shares.

The interest paid (dividends) by the firm is subject to the uncertainty of future earnings, plant investment, and management practices. In absence of precise information, one has to estimate both the market value and the predicted dividends (which likely will vary as a function of time) in order to estimate the cost of capital for issuance of common stock. It should be noted that in the calculation of the cost of capital for both preferred and common stock, the expenses (including those of the investment banker) must be recognized.

## 4.18  Summary

It is a rare event to find a contractor who is totally self-sufficient on his own funds for the operation of his firm. More than likely, the firm will find that at some point in time it is in need of financing for use as working capital or for the purchase of equipment.

Financing arrangements can be classified as short-term, intermediate-term, or long-term. The contractor is usually involved in short-term and intermediate-term financing. This is usually in the form of the use of trade credit, loans from banks, or equipment finance arrangements from finance companies.

Long-term financing by means of the issuance of bonds and stocks is seldom used by contractors. This is especially true in the case of the small contractor. His lack of size and the nature of his business operations are usually not consistent with long-term funding agreements.

The contractor, in seeking funding, should be able to compute the effective interest rate associated with the using of the funds. The effective interest rate often differs from the stated interest rate because of discounts, required compensating balances, and installment agreements. In addition to the cost of capital implied in a funding agreement, the firm has to be aware of the agreement's affect on its management practices and ownership of its assets. A good funding arrangement is the result of much searching and negotiations.

# Chapter 5

# Financial Accounting

## 5.1 Introduction

There is probably no type of firm that needs sound accounting practices more than the construction firm. The success of the firm is closely tied to its ability to forecast and control costs. Both of these functions have accounting as their base. The lack of proper accounting practices is high on the list of reasons for construction firm financial failures.

There are several characteristics of the construction firm, to include the small and large firm, that result in construction accounting practices varying significantly from those of other industries. Included in these unique characteristics are the following:

- Construction projects are usually built away from the firm's main office; the result is that the accounting source documents are sometimes late or inaccurate.
- Usually each project is somewhat unique in regard to the design and construction process.
- Projects often overlap the firm's accounting reporting period; the result is that profit recognition is difficult and uncertain.
- The firm is very dependent on the financial reporting required of many external entities to include the surety company and lending institutions.
- The majority of contracting firms are small with few owners and improper accounting controls that result in a lack of checks and balances on accounting procedures.

Each of these somewhat unique characteristics dictates unique construction accounting practices, procedures, and considerations.

Accounting includes financial accounting, managerial accounting, auditing, and tax. Financial accounting remains the center of each of the unique practices. Financial accounting's objective is, in great part, prepar-

75

ing financial statements. These include the income statement and balance sheet or statement of financial position.

Financial statements are prepared to satisfy the needs of parties external to the contractor as well as to serve the needs of the owner of the firm. Among the external parties that may require financial data regarding the contracting firm are the following:

- Lenders
- Sureties
- Public agencies
- Clients
- Suppliers
- Employees

Fundamental financial accounting to include the financial transactions and the preparation of the financial statements are covered in this chapter. It is essential that the contracting firm have an understanding of this process

The entire financial accounting process is aimed at summarizing and classifying financial data. The financial data is commonly classified into common accounts such as cash, receivables, revenue, expense, etc. Whereas some of these elements or accounts relate to a company's income statement, others relate to its statement of financial position or balance sheet.

The process by which financial transactions of a firm are summarized and classified into individual accounts is part of what might be referred to as the accounting cycle. An overview of this cycle is illustrated in Figure 5.1.

The recognition of financial transactions is fundamental to the practice of accounting. Every firm engages in financial transactions daily; it makes sales, purchases material, pays employees, borrows money, and buys or leases equipment. A financial transaction can be thought of as a financial event that results in changes to the firm's financial structure. In turn, the financial structure is characterized by its assets, liabilities, equity, revenues, and expenses. These terms will be described later in this chapter.

Each financial transaction is either external or internal to the construction firm. Examples of external financial transactions include purchasing concrete from a vendor, paying a craftsman's wages, and purchasing a paver. Recognition of depreciation as an expense of owning construction equipment is an example of a financial transaction internal to the firm.

Regardless of whether it is external or internal, each and every financial transaction is recognized in a firm's accounting system. The number and type of transactions that occur are dependent on many factors, including the size of the firm, the type of work it performs, and its legal and financial structure. Typical financial transactions and their accounting treatments are illustrated in this chapter.

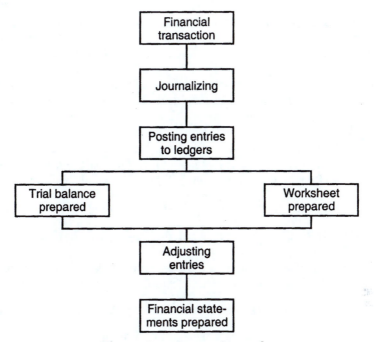

**Figure 5.1.** Accounting cycle.

## 5.2 Basic Accounting Principles

The treatment of a specific financial transaction in the accounting system is in part the result of several widely recognized generally accepted accounting principles. These principles have been issued by the American Institute of Certified Public Accountants (AICPA) and other accounting associations.

The *cost principle* indicates that cost is to be the measure of value assigned to the financial transaction. The contracting firm, even if it has purchased material or equipment at a bargain price, must recognize the transaction at cost. This removes subjective treatment by different firms. However, the cost principle may also result in a conservative recognition of a transaction's value and, in turn, a conservative view of the firm itself.

The *revenue principle* indicates when a transaction is to be recognized. It establishes different alternatives as to time of recognition. Included in these are the sales basis, the cash basis, the accrual basis, and the production basis. Each method can result in a transaction entering the accounting process at a different point in time. Whatever alternative is used, the *consistency principle* requires that the chosen method be applied the same at all times.

The *matching principle* emphasizes that financial transactions should be recognized in a manner that matches expenses with like revenues. Therefore, when feasible, expenses that are incurred in producing a product that yields revenue should be "matched" to the revenue in the period the revenue is earned.

These accounting principles, along with others, are very pragmatic and must be applied to diverse sets of facts and conditions. In view of these considerations, some exceptions are to be expected. These are part of the *exception principle*. The exception principle is especially relevant to the construction industry in view of the industry' unique considerations. An example of the application of the exception principle is the construction industry's use of the long-term revenue recognition methods, which include the percentage of completion method.

## 5.3   The Basic Accounting Relationship

An understanding of the accounting process and financial transactions will be achieved if the following basic relationship is kept in mind:

$$\text{Assets} = \text{Liabilities} + \text{Net worth} + \text{Revenue} - \text{Expense}$$

This equality not only prevails at the end of an accounting period, it prevails continuously. After each financial transaction, the five accounting categories — assets, liabilities, net worth, revenue, and expense — must be related using the above equality.

A company's assets may be viewed as all the things that add to the company's value. The liabilities of a company refer to its debts, or all the things it owes. The net worth, sometimes referred to as equity, is the amount of money invested in the firm and the profits retained.

Information regarding the contracting firm's assets, liabilities, and net worth is presented in the firm's balance sheet (also referred to as its statement of financial position), which represents the firm's condition at a given date.

Revenue represents the amount of money taken in by the company. An expense is the money spent for things that do not add directly to the company's assets. For example, payment of a utility bill would be an expense as it does not add directly to the company's assets.

Information regarding the contracting firm's revenues and expenses is presented in the firm's profit and loss statement (also referred to as its income statement). This statement illustrates whether the company derives a profit or incurs a loss for a specified period of time.

Let us illustrate the equality of assets, liabilities, net worth, revenue, and expense in terms of a particular construction firm. Let us assume we invest $40,000 in the formation of a contracting firm. We have then created assets of $40,000 in the company. The accounting equality is as follows:

$$\text{Assets} = \text{Net worth}$$
$$\$40,000 = \$40,000$$

Let us now assume that we buy \$30,000 worth of new construction equipment for our company. Rather than pay cash for the equipment we buy it on credit. We have now increased the assets of our company by \$30,000. However, we have created a company liability of \$30,000 (we eventually will have to pay for the equipment). The accounting equality is as follows:

$$\text{Assets} = \text{Liabilities} + \text{Net worth}$$
$$\$70,000 = \$30,000 \quad + \$40,000$$

Let us assume that our newly formed contracting firm receives a contract for building a construction project. While building the project we have to pay our employees. Let us assume that at the end of a particular week we pay them a total of \$3,000. Let us assume that we pay them cash, reducing our company assets by \$3,000. We have incurred an expense of \$3,000 which has not directly increased our company's assets. The accounting equality is as follows:

$$\text{Assets} = \text{Liabilities} + \text{Net worth} - \text{Expense}$$
$$\$67,000 = \$30,000 \quad + \$40,000 \quad - \$3,000$$

At the end of the month, the project owner pays us \$4,000 for the work we have completed. This \$4,000 worth of cash increases our company's assets by \$4,000. Likewise, we have had a company revenue of \$4,000. The accounting equality now becomes the following:

$$\text{Assets} = \text{Liabilities} + \text{Net worth} + \text{Revenue} - \text{Expense}$$
$$\$71,000 = \$30,000 \quad + \$40,000 \quad + \$4,000 \quad - \$3,000$$

Obviously the contractor would have many more entries during its operation. Each entry would affect the accounting system. However, at all times the relationship would be an equality.

## 5.4 Debits and Credits

Accounting practice is to identify each element of the financial transaction as a debit or credit to the accounting equality. Each term of the accounting equality, including assets, liabilities, net worth, revenue, and expense, is divided into debits and credits.

The reader should be cautioned against attempting to relate any real significance to the words "debit" and "credit". Their significance is limited to the accounting process. There are two fundamental definitions that are part of recognizing financial transactions in the accounting equality: (1) an entry that increases the assets of the company is defined as a debit and is assigned a positive value; and (2) for every debit entry in the accounting equality, there must also be a corresponding credit entry.

As previously stated, the accounting financial equality, relating a company's assets, liabilities, net worth, revenue, and expenses, must balance after every financial entry. As defined, an asset debit is positive. For the relationship to remain an equality, and for consistency with the asset debit definition and an "equal debit–equal credit" definition, a debit is considered negative and a credit is considered positive if the entry is a liability, a net worth, or revenue. If an entry is an expense, it is positive if it is a debt and negative if it is a credit. Similarly, an asset credit is negative. The described convention of assigning positive or negative values to the various debts and credits in the accounting equality is summarized below:

| ASSETS | | = | LIABILITIES | | + | NEW WORTH | | + | REVENUE | | − | EXPENSE | |
|---|---|---|---|---|---|---|---|---|---|---|---|---|---|
| Debit | Credit | | Debit | Credit | | Debit | Credit | | Debit | Credit | | Debit | Credit |
| + | − | | − | + | | − | + | | − | + | | + | − |

Having included debits and credits in the accounting financial equality, let us now view the equality with respect to the formation of our previously discussed contracting firm. Originally we invested $40,000 in the formation of the company. The $40,000 increased our company's assets. Thus, the $40,000 is entered as a debit in the asset term of the accounting equality. We, as the investors, have created a net worth of $40,000. Consistent with the accounting model definitions, this becomes a credit entry in the net worth term of the accounting model. The accounting equality is as follows:

| ASSETS | | = | LIABILITIES | | + | NEW WORTH | | + | REVENUE | | − | EXPENSE | |
|---|---|---|---|---|---|---|---|---|---|---|---|---|---|
| Debit | Credit | | Debit | Credit | | Debit | Credit | | Debit | Credit | | Debit | Credit |
| + | − | | − | + | | − | + | | − | + | | + | − |
| 40,000 | | | | | | | 40,000 | | | | | | |

Having created the company, we purchased on credit $30,000 worth of new construction equipment. The new construction equipment increased our assets by $30,000. Thus, it is entered as a debit in the asset term of the accounting equality. The $30,000 we owe to the seller of the equipment is a liability. Since the equipment was assigned to "debit" in the asset term of the accounting equality, the newly created liability is a credit. The accounting equality is now as follows:

| ASSETS | | = | LIABILITIES | | + | NEW WORTH | | + | REVENUE | | − | EXPENSE | |
|---|---|---|---|---|---|---|---|---|---|---|---|---|---|
| Debit | Credit | | Debit | Credit | | Debit | Credit | | Debit | Credit | | Debit | Credit |
| + | − | | − | + | | − | + | | − | + | | + | − |
| 40,000 | | | | 30,000 | | | 40,000 | | | | | | |
| 30,000 | | | | | | | | | | | | | |

We then performed part of the work on a construction project and paid our workers a sum of $3,000. The payment of the $3,000 decreased our cash and, therefore, our assets. Thus the $3,000 is a credit in the asset term of the accounting equality. The $3,000 was a project expense. Consistent with the fact that the $3,000 was a credit in the asset term of the accounting equality, it is a debit in the expense term. The accounting equality is now as follows:

| ASSETS | | = | LIABILITIES | | + | NEW WORTH | | + | REVENUE | | − | EXPENSE | |
|---|---|---|---|---|---|---|---|---|---|---|---|---|---|
| Debit | Credit | | Debit | Credit | | Debit | Credit | | Debit | Credit | | Debit | Credit |
| + | − | | − | + | | − | + | | − | + | | + | − |
| 40,000 | 3,000 | | | 30,000 | | | 40,000 | | | 4,000 | | | 3,000 | |
| 30,000 | | | | | | | | | | | | | |

Finally, we received $4,000 revenue from the construction project owner. The $4,000 increased our company assets by $4,000. Thus, it is a debit. Similarly, the $4,000 is identified as revenue since it was received as a result of our project performance. Because the $4,000 was an asset debit, it is a credit in the revenue term of the accounting equality. The accounting equality is now complete and is as follows:

| ASSETS | | = | LIABILITIES | | + | NEW WORTH | | + | REVENUE | | − | EXPENSE | |
|---|---|---|---|---|---|---|---|---|---|---|---|---|---|
| Debit | Credit | | Debit | Credit | | Debit | Credit | | Debit | Credit | | Debit | Credit |
| + | − | | − | + | | − | + | | − | + | | + | − |
| 40,000 | 3,000 | | | 30,000 | | | 40,000 | | | 4,000 | | | 3,000 | |
| 30,000 | | | | | | | | | | | | | |
| 4,000 | | | | | | | | | | | | | |

Other financial transactions could be entered into the accounting equality. In actuality, the construction firm does not enter its transactions into the described accounting equality as presented. The accounting entries are made in journals. However, these journals and supporting ledgers are based on the described accounting equality.

## 5.5   Typical Construction Financial Transactions

The contracting firm engages in many types of financial transactions. Although the type and size of the construction firm plays an important role in characterizing the number and types of transactions, even the relatively small-sized contracting firm takes part in many daily transactions. Material purchasing, labor payment, equipment purchasing, and billings to clients necessitate numerous transactions.

The complexity of discussing typical construction industry financial transactions is increased by the fact that the firm may recognize transactions using numerous variations of alternative, accepted accounting methods. However, basically, the firm may use either a cash method or an accrual method.

## 5.6   Initial Investment

Every contracting firm starts with an initial investment of monetary and non-monetary assets by investors or owners of the firm. Although this initial investment may be very simple as to the accounting recognition (for example, the initial investment of $5,000 by a single investor in his own company), the initial transaction may also be very complex. For example, complex partnership investments and corporation stock issues may require several accounting journal entries. However, these complex investment structures are rare in the construction industry and are therefore not discussed in this book.

A large percentage of small contracting firms originate with the single investment by one or two individuals in the firm. This investment may be cash, equipment, or cash and equipment. The cash and equipment are considered assets of the newly created firm and the investment is recognized by creating a contributed capital investment account. For example, let us assume that Andy Jackson invests his life savings of $10,000 and his pickup truck in a newly created firm. Assuming the truck has a fair market value of $5,000 for an estimated remaining life of five years and is to be depreciated using the straight-line method, the initial accounting entry would be recorded as follows:

| | | |
|---|---|---|
| Cash | $10,000 | |
| Truck (5-year life, straight line) | 5,000 | |
| Andy Jackson, proprietor | | $15,000 |

It should be noted that the debit entries (those listed first) are equal in amount to the credit entry. Andy Jackson's contributed capital account (referred to as his equity or capital account) is increased when he makes another investment or may be decreased if he decides to make a withdrawal of the invested capital.

Let us assume that instead of Andy Jackson's sole investment, he joins forces with his friend Steve Janko in forming a partnership to start a small contracting firm. Andy Jackson's initial contribution is still $10,000 cash and $5,000 for his fair market value truck. Steve Janko's investment is $5,000 cash. Assuming Jackson and Janko are to be credited for their investments, the initial accounting entry would be as follows:

| | | |
|---|---|---|
| Cash | $15,000 | |
| Truck (5 year life, straight-line) | 5,000 | |
| Andy Jackson | | $15,000 |
| Steve Janko | | 5,000 |

As an alternative to the described partnership agreement, let us assume that Steve Janko has a college education as an engineer and, as such, negotiates with Jackson to own a share of the business that is larger than his contribution. In particular, let us assume that the partnership agreement is such that Janko is to be credited with 50 percent of the firm's contributed capital.

It should be noted that Janko only invests 25 percent of the cash and equipment assets. How then is the accounting entry to be made to recognize a 50 percent ownership? This may be done by one of two alternative accounting entries. For one, the entry may be made using the bonus method; Jackson's capital account is reduced and the amount of reduction is added to Janko's account. The total capital remains unchanged, that is $20,000.

If Janko is to be credited for 50 percent of the total capital, his final capital account balance should be $10,000. However, he only contributed $5,000. As such, Jackson in effect gives up $5,000 of his investment and transfers it to Janko. The entry would be made as follows:

| | | |
|---|---|---|
| Cash | $15,000 | |
| Truck (5 year life, straight line) | 5,000 | |
|    Andy Jackson | | $10,000 |
|    Steve Janko | | 10,000 |

Had Janko contributed more than 50 percent of the total capital and was to be credited for 50 percent, Jackson, instead of Janko, would receive the bonus.

Instead of Jackson offering Janko a bonus, let us assume that in their agreement they decide that Janko's education is an asset to the firm and they want to treat it as such. This is done by the recognition of goodwill. In the goodwill method, an asset account labeled "goodwill" is debited for the value assigned to Janko's education. In the example in question, the dollar amount of goodwill to be recognized is determined by the fact that Janko's total capital account is to be 50 percent of the total capital. Jackson's capital account remains unaffected at $15,000. Thus, if Janko is to be credited with an equal amount, goodwill recognized must equal $10,000 and Janko's capital account increased to $15,000. The entry follows:

| | | |
|---|---|---|
| Cash | $15,000 | |
| Truck (5-year straight line) | 5,000 | |
| Goodwill | 10,000 | |
|    Andy Jackson | | $15,000 |
|    Steve Janko | | 15,000 |

Unlike cash, which is considered a current asset, and the truck, which is considered a fixed asset, the goodwill is an intangible asset. Such an asset has no physical value, but adds to the earning potential of the firm. Other types of intangible assets include patents and copyrights. Similar to fixed assets, intangible assets are not to remain indefinitely on the firm's financial statements. In particular, intangible assets are to be amortized (part of the value written off as an expense each accounting period). Goodwill is to be written off by consistent charges to an expense account over a period of forty years or less. Assigning goodwill to the "purchase" of an individual's skills is a questionable point. In general, the skills of employees and owners of a firm are not recognized as assets by a firm. For example, a construction firm with well-trained and experienced management certainly has an "asset" in that they add "value" to the firm. However, current accounting practice has not accepted human resource accounting.

The initial entry necessary to recognize the formation of a corporation varies somewhat depending on the complexity and number of shareholders. Generally, common stock is issued for the starting of a corporation. The common stock may be par-value or no par-value stock. Par-value stock has a designated dollar "value" associated with it. Normally the par value is low and the dollar amount receive for the stock exceeds the stated par value. The amount received in excess of the par value is referred to as premium-on-capital stock and is recognized separately from the dollar amount of the par value of the stock. Even if stock is no par-value stock, it is normally assigned a stated value. As such, it differs little from par-value stock. The par value of stated value of stock has little significance other than it is accounted for separately in the accounting process and some state statutes require a firm to retain the dollar amount of the par-value or stated-value stock. That is, the firm is prohibited from using this dollar amount for paying dividends or buying back issued shares of stock. This dollar amount of required capital to be retained is referred to as legal capital. A typical corporate stock issuance is recognized as follows:

| | | |
|---|---|---|
| Cash | $25,000 | |
| Common stock | | $20,000 |
| (10,000 shares at $2 par value) | | |
| Premium on common stock | | 5,000 |

## 5.7  Debt Capital

Not every contracting firm obtains monetary and non-monetary assets by means of contributed capital. The amount of assets required to start a sizable firm may exceed the amount of available contributed capital from owners of the firm. Even when equity capital can be obtained, the firm may choose to obtain some of its required assets by means of taking on debt. Such a practice can lead to favorable leverage, that is, higher earnings to investment can be obtained.

Purchasing assets on credit, signing a note payable, financing equipment through a lease (discussed in a following section), and issuing bonds are all examples of obtaining assets by means of debt capital. Issuing bonds as a means of obtaining debt capital is not a common construction industry practice in that this form of raising debt capital is generally available only to large firms. Regardless of whether the contracting firm is a proprietorship, a partnership, or a corporation, the required accounting entries for debt capital transactions are the same.

Obtaining debt capital creates a liability to the firm in that the capital has to be repaid (usually with an interest cost) to the lender. Liabilities are normally classified as current or long-term. A current liability is designated as an obligation whose liquidation is reasonable expected to require the use of existing resources properly classified as current assets or the creation of other current liabilities. Perhaps more simply stated, liabilities are debts the firm expects to repay within its operating cycle which normally can be taken as one year. Typical current liabilities are:

1. Accounts payable
2. Short-term notes payable
3. Advances held as returnable deposits
4. Bonus obligations

Most current liabilities are not normally recognized in a cash accounting method. As such, typical journal entries are absent in accounting for current liabilities when using the cash method.

The most commonly recognized liability for the construction firm is the purchase of material or supplies on credit. Assuming a purchase of $1,000 of materials or supplies for a given contracting firm's inventory, the accounting entry would be as follows:

| | | |
|---|---|---|
| Materials — inventory | $1,000 | |
| Accounts payable | | $1,000 |

The recognition of short-term notes, advances, taxes, or bonus obligations as a liability would result in an entry similar to the one above.

Because issuing bonds is not a common occurrence for the smaller or middle-sized construction firm, only basic entries for this issuance are discussed. Let us assume that, in order to expand its operations, a relatively large contracting firm decides to issue $100,000 in bonds payable in ten years with an annual interest rate of 5 percent. In other words, the contracting firm is willing to pay the purchasers of the bonds $5,000 interest each year and repay the $100,000 face value of the bonds after ten years. Assuming that the firm was successful in selling the bonds for exactly $100,000, the accounting journal entry necessary in the firm's books would be as follows:

| | | |
|---|---|---|
| Cash | $100,000 | |
| Bonds payable | | $100,000 |

The annual interest expense to the contracting firm would be recognized by the following entry.

| Bond interest expense | $5,000 | |
|---|---|---|
| Cash | | $5,000 |

The above entries assume that the firm would receive exactly $100,000 for the $100,000 face-value bonds. More likely than not, the firm would receive something less or more than the $100,000. If the firm's financial status were sound and the current interest rate for marketable securities were lower than 5 percent, the bond investor would probably be willing to pay more than $100,000 for the bonds. For example, the contracting firm might receive $105,000. The $5,000 is referred to as a premium on the bonds. The fact that the bonds sell at a price greater than the face value results in an effective interest rate (referred to as the yield) less than the stated face-value interest rate.

More likely than not, the contracting firm will receive less than the expected $100,000, assuming that the 5 percent is the current marketable interest rate. This is because the typical investor views the construction industry as a financially risky investment. Unfortunately, statistics prove the investor correct.

Let us assume that the contracting firm receives $95,000 from the issuance of the 5 percent $100,000 bonds. The $5,000 shortage is referred to as a discount on bonds. In effect, the firm is paying a higher interest rate than 5 percent in that it only receives $95,000 instead of $100,000. This higher interest rate can be calculated using yield tables.

Let us arrange this latter transaction in the required journal entries. The issuance of the bonds would be recorded as follows:

| Cash | $95,000 | |
|---|---|---|
| Discount on bonds payable | 5,000 | |
| Bonds payable | | $100,000 |

The bond discount account would be amortized along with the annual interest expense journal entry. Assuming a ten-year period of amortization, this would be as follows:

| Bond interest expense | $5,500 | |
|---|---|---|
| Cash | | $5,000 |
| Discount on bonds payable | | 500 |

In effect, the annual interest expense is increased $500. The above entry has been made assuming straight line amortization of the discount. The accounting entries required for handling the premium on bonds would be similar to those made for the discount on bonds.

## 5.8   Accounting for Cash

Cash is considered a current asset in that a firm will use the cash throughout its operating cycle. Normally, cash is made available to the firm from initial investment (equity investment or from creditors of debt), or from the income from operations. Less frequent in occurrence is obtaining cash from selling one's assets or from selling a portion of the business. Cash obtained as interest from temporary investment of cash (for example, savings accounts, short-term securities, and so on) is debited along with a credit to interest income.

Accounting for cash normally presents little difficulty. There are no estimates required or valuation problems in that the accounting unit of measure is expressed in the monetary unit. The main difficulty in accounting for cash is to provide a means of controlling cash receipts, payments, and petty cash.

The numerous daily cash transactions that are part of running a contracting firm and construction project necessitate the existence of one or more petty cash funds. Items such as small hand tools, paper cups, refreshments, and miscellaneous supplies may be purchased from cash in the petty cash fund. Because of the availability of the cash in this type of fund, there has to be concern in regard to controlling cash payments from it.

Two different systems can be used to control a petty cash fund. One system is referred to as a fluctuating-fund system. The initial accounting entry is the same for each of the petty cash systems. Let us assume a firm creates a $500 petty cash fund. The initial accounting entry is as follows:

| | | |
|---|---|---|
| Petty cash | $500 | |
| Cash | | $500 |

The petty cash account, like the cash account, is considered an asset to the firm. This transaction merely transfers funds from "normal" cash to petty cash.

Accounting control and entries for transactions after the initial transaction vary depending on the system used. Generally, when a fluctuating system is used, an individual in control of the fund is required to maintain a petty cash book. Disbursements are entered in a book that is used as a basic record when the fund is replenished. As disbursements are made, an expense account is debited and the petty cash fund is credited. For example, let us assume that a few small tools costing $40 are purchased from the petty cash account. The accounting entry would be as follows:

| | | |
|---|---|---|
| Small tools expense | $40 | |
| Petty cash | | $40 |

After a period of time the petty cash fund is replenished by a debit to petty cash and a credit to cash. However, this additional cash is not necessarily equal to the cash disbursements made. It is evidence that, when using this approach, the balance in the petty cash account fluctuates.

The non-fluctuation, or imprest petty cash system is normally preferred in that it is easier to audit than the fluctuating system. Using the imprest system, disbursements are made from the petty cash fund and

recorded and summarized in a petty cash book. No accounting debit or
credit is made to adjust the initial petty cash entry unless it is desired to
increase or decrease the fund. Instead, after the petty cash book is summarized, a check is drawn to the petty cash fund for the exact amount of
the expenses. A journal entry is made that debits the appropriate expense
accounts and credits cash. For example, let us assume that, during a given
time period, the only disbursement from the petty cash fund was the previously described $40 small tools expense. Using the imprest-fund system,
the required accounting entry would be as follows:

| | | |
|---|---|---|
| Small tools expense | $40 | |
| Cash | | $40 |

Other petty cash disbursements would be handled in a similar manner.
As can be observed, the petty cash fund does not fluctuate.

Regardless of whether the fluctuating or imprest system is used, the
firm should immediately investigate any abnormal overages or shortages.
Unfortunately, overages can usually be explained whereas shortages may be
difficult to pinpoint and correct. Overages or shortages may be identified
by means of a bank reconciliation of cash.

## 5.9   Accounting for Fixed Assets

The typical contracting firm has a substantial share of its contributed capital and the debt capital invested in fixed assets. Fixed assets are properties and rights that the business retains more or less permanently to use
in performing its production operation. Fixed assets such as plant (office
buildings, furniture, etc.) and equipment may make up a large percentage
of some contracting firm's assets. Construction equipment in particular
plays an important role in the production process performed by the typical
contracting company. Pavers, cranes, reusable forms, trucks, and concrete-
mixing equipment are just a few of the types of equipment used by the
contracting firm. Construction equipment is usually recognized by three
different classifications: (1) heavy equipment, (2) miscellaneous tools and
equipment, and (3) trucks and autos. The large dollar value associated with
such equipment has an important impact on the firm's financial statements.

The contracting firm normally has several options available to it in
regard to the use of construction equipment. It may choose to purchase a
piece of equipment, or, as an alternative to purchasing a piece of equipment,
the firm may choose to rent or lease it. Several different types of lease
agreements may be arranged each of which requires different accounting
treatment and each of which has a different impact on the firm's financial
statements.

Let us assume a contracting firm purchases a piece of equipment. The
equipment should be recorded at an asset value equivalent to purchase cost

plus any shipment or installation costs. Available discounts, whether taken or not, should be deducted from the equipment's invoice cost. For example, let us assume a firm purchases a $40,000 machine and the manufacturer charges an additional $200 for transporting the equipment. The agreement with the contractor firm indicates it will receive a 5 percent discount on the $40,000 if cash is paid when the machine is delivered. However, the contracting firm only pays $10,000 upon delivery with the balance due covered by a note payable which is given to the equipment dealer. The journal entry required to recognize the purchase of the machine is as follows:

| | | |
|---|---|---|
| Equipment — machine | $38,200 | |
| Allowance for discount | 2,000 | |
| Cash | | $10,000 |
| Note payable (8% annual interest) | | 30,200 |

Thus, the machine is capitalized at $38,200. It should be observed that the potential $2,000 discount has been deducted from the $40,000 invoice cost. The lost discount would be recognized as an expense to the contracting firm. The journal entry necessary to recognize the expense is as follows:

| | | |
|---|---|---|
| Discount lost expense | $2,000 | |
| Allowance for discount | | $2,000 |

Rather than purchase a piece of construction equipment outright, the firm may purchase a piece of equipment by "trading in" an old piece of equipment as part payment for the new equipment. The accounting cost principle requires that the new equipment be valued (i.e., capitalized) as the sum of the fair market value of the old equipment traded in plus any cash given to the seller. Any difference between the book value of the old equipment and its par market value at the date of exchange should be recognized as a loss or gain on the disposition of the equipment.

Let us assume a contracting firm trades in an old truck for a new one. The old truck's original cost was $20,000; depreciation to date on the truck is $15,000 (leaving a book value of $5,000). The new truck has a selling price of $25,000. The contracting firm traded the old truck and $17,000 cash for the new truck. Since the firm paid $8,000 less cash than the selling price of the new truck, one can assume that the old truck has a fair market value of $8,000. As such, the accounting entry is made as follows:

| | | |
|---|---|---|
| Truck — new ($17,000 + $8,000) | $25,000 | |
| Accumulated depreciation — old truck | 15,000 | |
| Cash | | $17,000 |
| Truck — old | | 20,000 |
| Gain on exchange of trucks | | 3,000 |

The $3,000 gain is reported on the firm's income statement. Similarly, any loss on a similar exchange would be handled in a similar manner. The debit to "accumulated depreciation" is required to offset earlier credits to this account.

Because of the type of work the contracting firm performs, it is not unusual for the typical firm to fabricate some of its own equipment. The accounting problem of recognizing general overhead arises when a firm constructs its own assets. Two possibilities exist. For one, the contracting firm may construct its equipment when in fact it is operating at capacity. That is, all of its personnel are gainfully employed. In this situation, clearly the asset constructed should be charged for additional overhead costs and a share of the firm's general overhead.

Equipment repairs and maintenance are a normal occurrence when owning construction equipment. The relevant question in regard to accounting for repairs and maintenance is whether such costs should be capitalized (deducted from income over a period of years) as part of the equipment cost or expensed as a period cost in one year. In general, only those expenditures that add to the production capacity of the equipment should be capitalized. Costs that are expended to maintain the equipment's production level, or to restore the equipment to its normal production level, should be expensed. The end result is the most, if not all, construction equipment repairs and maintenance costs should be expensed rather than capitalized.

## 5.10   Depreciation of Fixed Assets

Fixed assets, including construction equipment, are purchased because of their future revenue generating potential. As such, as the revenue is generated, a portion of the cost of the asset should be matched periodically with the period revenue. This process is referred to as depreciation accounting. In particular, depreciation accounting is a system of accounting that aims to distribute the cost of a tangible capital asset, less any salvage, over the estimate useful life of the units in a systematic and a rational manner. Depreciation accounting is a process of allocation, not of valuation.

Note should be made of the fact that depreciation must be recognized in a systematic and rational manner. Alternative depreciation methods such as the straight-line method, the declining-balance method, and the sum-of-the-digits method have all been judged systematic and rational. It is possible to use one depreciation method for financial accounting and yet another for tax purposes.

In determining the periodic depreciation charge for a piece of equipment, the initial cost of the equipment, its salvage value, and the equipment's useful life must be determined. As to the useful life, it should be noted that contracting firms sometimes purchase equipment for a specific project. When the project is completed, the firm disposes of the equipment rather than retain it for further work.

Regardless of which depreciation method is used to depreciate a given piece of equipment, the type of accounting journal entry remains the same. For example, let us assume that, using an acceptable depreciation method, a periodic depreciation charge on a given truck is determined as $2,000. The periodic accounting journal entry would be as follows:

| | | |
|---|---|---|
| Equipment depreciation expense | $2,000 | |
| Accumulated depreciation — truck | | $2,000 |

The dollar value of the entry would depend on the cost of the asset, its useful life, the salvage value, and the depreciation method used.

The firm should use a depreciation method that best matches the utilization of the equipment to the revenue earned. However, the practicality of a given method overrides the matching concept. Several depreciation methods determine the deprecation charge as a function of time. Others determine the depreciation charge as a function of asset usage.

## 5.11  Accounting for Leases

The large dollar cost associated with the purchase of equipment and the rather specialized characteristics of some types of construction equipment result in a firm renting or leasing equipment used to build projects. In addition to having less initial negative effect on the firm's cash flow (its liquidity), renting or leasing equipment can result in a smaller unit cost of usage when the firm only needs the equipment for a few select operations, whereas it would remain idle for long periods if it were owned. On the other hand, if equipment is used rather continuously by the construction firm, purchasing the equipment normally results in a much lower cost per unit of usage. However, even when this is the case, the construction firm's cash availability may restrict the firm to renting or leasing a needed piece of equipment.

When equipment is rented, the accounting is relatively simple. The rental cost of the equipment is allocated to particular projects on some reasonable basis such as direct cost of the projects, duration of the projects, equipment time on the projects, or equipment mileage on the projects. The accounting entry to record the allocated cost is made by debiting a project expense entry to record the allocated cost and crediting cash paid for the rental or an account payable if the rental company issues the equipment on credit. Assuming a $2,000 rental cost is paid by a contracting firm for the use of a piece of equipment on a single work item, the accounting entry would be as follows:

| | | |
|---|---|---|
| Project expense — equipment rental | $2,000 | |
| Cash | | $2,000 |

Unlike purchased equipment, equipment rentals do not appear as an asset on the firm's financial statements. However, substantial rental commitments should be disclosed in footnotes to a contracting firm's financial statements.

Leasing differs from renting in that leases are normally considered to be for a longer period of time. For example whereas a contracting firm may rent a piece of equipment for a one-month duration, a lease agreement for the equipment might run for years. More important than the duration difference between leasing and renting is the fact that leases may in fact be purchases or options to purchase. Such agreements require accounting recognition that differs from that used for recognition of rental agreements.

Under some circumstances, a lease agreement may actually represent an installment purchase of equipment. While the recognition of such an agreement may sometimes be difficult to detect, it is normally characterized by the following circumstances.

1. When the lease is made subject to the purchase of the equipment at the expiration of the lease for a nominal sum or for an amount obviously much less than its fair value at the time of purchase.
2. When the lease agreement stipulates that the rentals may be applied in part as installments on the purchase price of the equipment
3. When the rentals obviously are not comparable with other rentals for similar equipment so as to create the presumption that portions of such rentals are partial payments under a purchase plan.

Additional means of determining whether an agreement is a lease or an installment purchase should focus on which party is liable for maintaining the equipment and which party bears the risk of ownership.

In a true lease (sometimes referred to as an operating lease) the owner (the lessor) retains the usual risk of ownership. The lessee's rental receipts are computed to cover the usual ownership cost of depreciation, taxes, insurance, and profit. The lease is normally readily cancelable by the less or, and no special equipment ownership is given up by the lessor.

Typically the contracting firm would prefer to recognize a piece of equipment as a lease rather than a purchase. The purchase requires recognition of a substantial liability. A bonding company or lending institution may look unfavorably at this liability on the firm's balance sheet.

The contracting firm's accounting recognition of an operating lease is similar to the recognition of a rental agreement. That is, a debit to rent or lease expense and a credit to cash is recognized. The entry is exactly the same as the one previously made for the described rental payment.

If the firm prepays some of the equipment lease expense (as is often required), that portion which is prepaid should be recognized as an asset rather than as an expense in the period paid. For example, let us assume a construction firm initially pays an equipment leasing firm $4,000 for the lease of a piece of equipment, $2,000 of which is to cover the lease cost in the current period. The accounting journal entry would be as follows:

| | | |
|---|---|---|
| Lease expense | $2,000 | |
| Prepaid lease expense | 2,000 | |
| Cash | | $4,000 |

The $2,000 prepaid lease expense amount would be removed in the following accounting period and an expense recognized.

## 5.12   Revenue and Expense Recognition

Revenue and expense transactions are handled through nominal accounts. Nominal accounts are those accounts whose balances are normally transferred (i.e., closed) to other accounts when the books are closed. In particular, revenue and expense accounts are closed to an account commonly referred to as income summary. In effect this account is used to summarize all the revenue and expense transactions for a given period of time. The difference between the two types of accounts is reported as profit or loss for the period.

The procedure as to how and when various construction revenue and expense transactions are entered and summarized in the firm's set of books is in part dependent on the firm's choice of accounting method. For example, the percentage-of-completion method and the completed-contract method differ as to when revenue and expenses are summarized into the income account.

Independent of the accounting method choice, certain fundamental revenue and expense transactions are common to each and every firm. Let us first consider common expense transactions.

The contracting firm may purchase materials for a specific project or may purchase the material and add it to its inventory. The former transaction is an expense transaction whereas the latter is merely an exchange of asset accounts. Let us consider the expense transaction.

If the firm pays $1,200 cash for the material the transaction would be as follows:

| | | |
|---|---|---|
| Material expense | $1,200 | |
| Cash | | $1,200 |

A firm utilizing a job-cost accounting system, which identifies material to specific projects, may describe the entry as follows:

| | | |
|---|---|---|
| Material expense — Job 101 | $1,200 | |
| Cash | | $1,200 |

A more sophisticated job-cost accounting system may identify material and labor costs to specific segments of work of a project. The following entry would reflect such recognition.

| Material expense — concrete slabs — Job 101 | $1,200 | |
| Cash | | $1,200 |

The latter two types of transactions would be kept in subsidiary job-cost ledgers and summarized via a general ledger material expense account.

More often than not material is purchased on account. In this case the credit entry would be to an accounts payable account rather than the cash account. Assuming material is traced to specific projects, the entry would be as follows:

| Material expense — Job 101 | $1,200 | |
| Accounts payable — Ace Hardware | | $1,200 |

When the invoice is paid, an entry is made to reflect the change in the asset and liability accounts. If the firm were on a cash basis accounting method, the expense would not be recognized until the material bill was paid.

Often the construction firm is given an opportunity to take advantage of a purchase discount when paying for material. For example, if the construction firm pays the material vendor within ten days, it may receive 2 percent discount from the invoice. The entire bill amount is due in thirty days. This payment agreement should be indicated by the notation 2/10/n30. The recognition of purchase discounts can be handled in more than one manner. The question arises as to whether the failure to take advantage of a purchase discount should create recognition of an expense entitled "purchase discount lost". Let us assume a construction firm has a potential 10 percent discount and fails to take advantage of it on a $1,200 invoice. One means of transaction recognition is as follows:

| Materials expense | $1,200 | |
| Cash | | $1,200 |

Alternatively, the lost purchase discount could be singled out through the following entry:

| Material expense | $1,080 | |
| Lost purchase discount expense | 120 | |
| Cash | | $1,200 |

The latter entry has the advantage of singling out the inefficiencies of the cash disbursement function. The expense "lost purchase discount" is identified as an "other expense" in the firm's financial statements.

If material is purchased for inventory, no expense account is immediately recognized. Assuming the material is purchased on account, the transaction would be as follows:

| Material inventory | $1,200 | |
| Accounts payable S & A Hardware | | $1,200 |

Eventually the material may be used for specific projects and an expense would be recognized:

| Material expense — Job 101 | $1,200 | |
|---|---|---|
| Material inventory | | $1,200 |

The transaction assumes the material is identified as belonging to a specific project rather than individual work segments.

Direct labor costs for a project, like material costs, may merely be identified as labor costs to the firm or may be identified as belonging to specific projects or segments of work. Labor expense entries must include recognition of withholding and FICA taxes. This may be done by means of two separate entries: one to reflect the labor wage expense, and the other to reflect various employer labor tax expenses.

The employer withholds employee taxes from the paychecks it issues. Let us assume that labor payroll is $10,000 for a certain week and employee payroll taxes withheld are equal to $1,800. The following entry would be made:

| Labor payroll — Job 101 | $10,000 | |
|---|---|---|
| Withholding taxes payable | | $1,200 |
| FICA taxes payable — employees | | 600 |
| Cash | | 8,200 |

In addition the employer would be liable for payroll taxes. Assuming they total $900 the entry would be as follows:

| Expense — payroll taxes | $900 | |
|---|---|---|
| FICA taxes payable — employer | | $600 |
| FUTA taxes payable — federal | | 50 |
| FUTA taxes payable — state | | 250 |

The FICA tax is social security and the FUTA is unemployment taxes. The employer will have to submit both the employee taxes withheld and its taxes payable to the appropriate agencies. However, when they are paid, an expense entry is not made. Rather the "accounts payable" liability accounts are debited and cash is credited for a like amount.

Supervision and administrative payroll entries would be made in a manner similar to the direct labor expense entries. However, rather than identify the expense as belonging to a specific project or work segment expense account, these labor accounts may be titled "officers salaries", "administrative salaries", and so on. Like direct labor costs, they will eventually be included in all the expenses that make up the firm's income statement.

Other types of expense transactions besides direct material and labor are recognized in the firm's income statements. The depreciation and lease expense entries were discussed in earlier sections. Additionally, the firm incurs many expense-related transactions that are not always easily traced to specific projects. They are commonly referred to as overhead expenses.

Using job-cost accounting (which is common to contractor) overhead is often expensed to specific projects through an allocation procedure. For example, after estimating total overhead expenses for a period of time, the firm may decide to apply overhead costs to projects at a rate of $1.50 per $1.00 of direct material and labor costs for a Project 101 for a period in time. The overhead expense entry would be made as follows:

| | | |
|---|---|---|
| Overhead applied expense — Job 101 | $7,500 | |
| Overhead control | | $7,500 |

The overhead control account is an account used to collect and record actual overhead expenditures. For example, assuming $500 of small tools is purchased (and identified as an overhead item), the purchase would be entered as follows:

| | | |
|---|---|---|
| Overhead control | $500 | |
| Cash | | $500 |

If overhead is being applied correctly, the debits and credits to the overhead control account should equal zero. Otherwise the firm has under- or over-applied overhead. This under- or over-applied overhead should be recognized through an adjustment to the expense accounts of the various projects.

As an alternative to applying overhead to specific projects through formal accounting entries, overhead costs may be merely expensed to specific overhead expense accounts as they are incurred. In this case, the specific account entry would depend on the type of overhead expense incurred. For example, if the overhead item were small tools, the entry would be as follows:

| | | |
|---|---|---|
| Small tools overhead expense | $500 | |
| Accounts payable — S & A Hardware | | $500 |

Other overhead expenses would be entered in a similar manner.

The point in time at which a revenue transaction is recognized depends on the choice of accounting method. Included in these methods are the cash and accrual methods. Using the cash method, the revenue accounting entry would be as follows:

| | | |
|---|---|---|
| Cash | $6,000 | |
| Revenue — Job 101 | | $6,000 |

More often than not, revenue is actually "earned" by the construction firm before the cash is received. The time of billing is often used as the point in time at which the revenue is recognized. The following entry would be made to reflect this occurrence:

| | | |
|---|---|---|
| Accounts receivable — Job 101 | $6,000 | |
| Revenue — Job 101 | | $6,000 |

When the cash is received, an entry is made to reflect the change in asset accounts. In particular, the entry would be made as follows:

| | | |
|---|---|---|
| Cash | $6,000 | |
| Accounts receivable — Job 101 | | $6,000 |

This transaction does not affect a revenue or expense account.

A reality of doing business is the fact that the firm may not be able to collect all of its receivables. In order to reflect this, a firm will recognize an expense entitled "bad debt expense". Several procedures for recognition of a bad debt expense exist. For one, the firm may write off the expenses as they are incurred. This is sometimes referred to as the direct write-off method. Assuming $500 of receivables were identified as uncollectible, the accounting entry would be as follows:

| | | |
|---|---|---|
| Bad debt expense | $500 | |
| Accounts receivable — Job 101 | | $500 |

Rather than utilizing the direct write-off approach, the contracting firm may recognize the bad debt expense by using an allowance account. Based on an estimate of monthly or yearly uncollectible receivables, the firm would set up an allowance account. Assuming for a period of time, say a year, that estimated uncollectible receivables are $4,000, the allowance entry would be as follows:

| | | |
|---|---|---|
| Bad debt expense | $4,000 | |
| Allowance for doubtful accounts | | $4,000 |

This entry recognizes the total estimated expense for the period and sets up a liability allowance account. As particular receivables are judged uncollectible, they are written off through the allowance account. Assuming a $400 receivable is judged uncollectible, the entry would be as follows:

| | | |
|---|---|---|
| Allowance for doubtful accounts | $400 | |
| Accounts receivable | | $400 |

Similar entries would be made throughout the year as other receivables were deemed uncollectible. If the initial estimate of the bad debt expense was correct, the allowance account will have a zero balance at year end. If the estimate was incorrect and thus a balance exists in the allowance account, the amount can be carried forward to the next period or adjusted to the year's bad debt expense account.

The contracting firm may receive revenue from sources other than normal operations. For example, it may receive revenue from interest or from rental of real property. It is common to recognize these non-operating revenues through "income" accounts. For example, assuming $300 interest is received, the entry may appear as follows:

| | | |
|---|---|---|
| Cash | $300 | |
| Interest income | | $300 |

Other non-operating revenues would be reflected in a similar manner. As with operating revenue, if recognition were to be identified before receipt of cash, an account receivable would be debited instead of the cash account.

## 5.13    Accounting Entries for Revenue Recognition

The choice of revenue-recognition method used by the contracting firm affects both the firm's financial statements and its calculated taxable income. Usually a method that is favorable for one is unfavorable for the other. For example, reducing reportable income by using an available revenue-recognition method may minimize the firm's tax liability. However, this reduction of income may also have a negative impact on the firm's financial statements including a smaller income, less net worth on the balance sheet, and probably less working capital on the balance sheet. Thus, while the objective of tax minimization might be achieved, the resulting minimization might be made at the expense of causing external parties such as lending institutions and sureties to view the firm's financial statements with disfavor.

Fortunately the above difficulty can be avoided. There is no requirement to use the same revenue-recognition accounting method for both financial reporting and determining tax liability. Often it is preferable to utilize separate methods for the two purposes.

## 5.14    Cash and Accrual Methods

A short-term construction project is usually considered one that has a duration of a year or less. Recording revenue and expense for short-term projects presents few problems relative to the recording needed for long-term projects.

Two methods of recognizing revenues and expenses for short-term construction projects are in general usage; these are the cash method and the accrual method. Some contracting firms have also devised "hybrid" methods for recognizing revenue that in effect combine elements of the cash and accrual methods.

When the cash method of accounting is used, revenue is recognized only when it is received or constructively received. (An example of revenue that is constructively received is interest posted to an individual's savings account that can be drawn upon at any time.) Using the cash method of revenue recognition, a construction firm would not recognize any revenue at the time it billed an owner for work performed. Revenue would be recognized only when the cash itself was received.

Expenses recognized using the cash method of accounting would be limited to those in which cash was actually expended. An exception to this would be non-cash expenditures creating an asset having a useful life that extends substantially beyond the close of a firm's operating cycle or a year. Examples of such non-cash expenditures that would be recognized in the cash method of accounting would be depreciation and an extended life insurance policy.

The accrual method of accounting recognizes revenue when events occur that fix the right to receive revenue and the amount thereof can be determined with reasonable accuracy. Therefore, when a contracting firm bills an owner for work performed, revenue is recognized at the time of the billing regardless of the date the amount is actually collected.

The accrual method recognizes expenses when events occur that establish the fact that the contracting firm has incurred a liability and the amount can be reasonably determined. For example, when a construction firm purchases material on account from a vendor for building a project, a liability is established; thus, the contracting firm recognizes an expense.

When the cash method of accounting is used, income (profit or loss) is calculated as the difference between cash collected and cash expended. Using the accrual method, income (profit and loss) is calculated as the difference between amounts billed to customers and expenditures regardless of whether the expenditures are paid or merely incurred.

Many accounting journal entries would be treated the same using either the cash or accrual accounting method. For example, as noted, depreciation expense would be recognized identically in both methods. However, if cash is not exchanged in a revenue or expense transaction, the accounting recognition would differ depending on whether the cash or accrual method was used.

Consider the purchase of $5,000 of material on account for building a project. If the accrual method were utilized, the journal entry would be

Project material expense                     $5,000
    Accounts payable                                         $5,000

When the vendor was paid the $5,000, the journal entry that would be made using the accrual method would be:

Accounts payable                             $5,000
    Cash                                                     $5,000

If the cash method were utilized, the journal entry for the material purchase would not be recognized at the time the material was purchased. Because the material is purchased on account and no cash was expended at that time, no accounting journal would be made. When the vendor bill was paid, the cash method of accounting entry would be made as follows:

Project material expense                     $5,000
    Cash                                                     $5,000

Note that, regardless of whether the cash or accrual method of accounting is used, the final impact on the contractor firm's books would be the same. That is, the firm's books would reflect an expense for material and a lessening of its cash account. While the final result may be the same, the interim results may be significantly different. For example, if an income statement were prepared after the material purchase but before the cash payment to the vendor, using the accrual method it would include an expense for material, whereas, using the cash method, it would not reflect the expense.

The cash method of revenue-recognition results in misleading financial statements. The failure to recognize expense incurred but not expense paid, such as the discussed expense for material, can result in a misleading or deceptive financial statement. For this reason, the cash method of accounting for financial reporting is not recommended. The cash method is usually only an excuse for poor bookkeeping that fails to recognize receivables and payables.

# 5.15    Effect on Income Statement of Cash and Accrual Methods

The cash or accrual methods of recognizing revenue and expense can result in widely varying recognized income depending on which method is used. If amounts billed and uncollected are in excess of costs and expenses incurred but not paid, the accrual method of accounting will result in more recognized income than the cash method.

In order to illustrate the differing impact of the cash and accrual methods on financial statements, let us consider the data for four different projects as illustrated in Figure 5.2. General and administrative (G & A) expense data are also shown in Figure 5.2. Note that the expenditures shown for each of the four projects are for direct job costs and do not include a portion of the general and administrative expense.

The impact of the cash versus accrual methods for the four projects in Figure 5.2 and the construction company as a whole is illustrated in Figure 5.3.

Naturally the results shown in Figure 5.3 are dependent on the data given. While using the cash method in Figure 5.3 has resulted in significantly more income than the accrual method, the reverse could also be true if different data had been given in Figure 5.2.

Independent of whether the cash or accrual method results in a larger income for a given set of data, note the significant potential difference in the reported income from the two methods illustrated in Figure 5.3. Especially noteworthy is the fact that the cash method can be interpreted as "reflecting" $90,000 of fictitious income in that outstanding payables exceed outstanding receivables by this amount. This $90,000 difference between payables and receivables is reflected in the income reported using the accrual method that is illustrated in Figure 5.3.

The failure to recognize payables and receivables has led many to conclude that the cash method is weak and non-representative of the firm's "actual position". This is true and especially important in regard to the construction firm in that there are usually a significant amount of payables and receivables.

It is also true that some individuals would argue that the accrual method is equally or more misleading than the cash method in regard to recognizing income. This argument also has merit; it acknowledges that contracting firms may "overbill" and therefore create unearned income or may receive invoices from vendors late, which has the effect of lessening the recorded expense. While both of these events are possible, the accrual method usually better reflects the actual income of the firm than does the cash method and is the preferred method of the two for financial reporting.

## 5.16  Long-Term Contracts

Because of the large dollar value of the construction project, the relatively complicated and time-consuming construction process, and the construction industry's dependence on weather and seasons, the duration of building a project often exceeds a year or overlaps the construction firm's year end. This, along with the fact that a contracting firm may overbill a customer for work performed as of a given date and the difficulty of determining the cost to complete a project, have led to the acceptability of long-term contract methods for recognizing revenue and expense in the construction industry.

The use of long-term contract accounting methods is somewhat unique to the construction industry because of the industry's unique characteristics. Included in long-term contract methods of revenue and expense recognition are the percentage-of-completion method and the completed-contract method.

The discussion of long-term revenue and expense-recognition methods is relevant to both financial reporting and calculating taxable income. Current financial statement presentation guidelines emphasize the use of the percentage-of-completion method. They argue that a contractor should almost always be able to make reasonably dependable estimates of the extent of project progress.

The procedures of using each of the methods will now be illustrated.

## 5.17  Percentage-of-Completion Method

The percentage-of-completion method is generally the preferable method for recording income on projects. Its major advantage is that it recognizes income on a current basis and, as such, results in a more regular flow of income. On the other hand, the method has the disadvantage of dependence on cost estimates, which are subject to uncertainties. The percent complete for an in process project is normally calculated one of two ways.

1. The percentage of total incurred costs to date divided by the estimated total costs after giving effect to estimates of costs to complete based upon most recent information.
2. The percentage of work completed that may be indicated by such other measures of progress toward completion as may be appropriate having due regard to work performed; for example, the square feet of concrete placed to date relative to the total required.

In most circumstances, the first means of estimating income to be recognized will prove appropriate. However, for certain types of projects total incurred costs may not be the best measure of income earned. Where a more meaningful allocation of income would result, the second means of estimating income is to be preferred. For example, in certain circumstances a more appropriate recognition of income may be based on the percentage of direct labor costs expended to the estimated total direct labor costs for the project.

Method 1 is the most commonly used for determining revenue, expenses, and income to recognize. To illustrate the calculation process, let us assume the following project data.

| | |
|---|---|
| Contract amount | $1,000,000 |
| Cost to date | 600,000 |
| Estimated additional costs | 200,000 |
| Total estimated costs | 800,000 |
| Billings to date | 800,000 |

The percentage of completion based on costs would be calculated as follows:

$$\text{Percent of Completion} = \frac{\text{Cost to Date}}{\text{Total Revised Est'd Cost}} = \frac{600,000}{800,000} = 75\%$$

The revenue to recognize would be calculated as follows:

(Contract amount) × (% Complete) = ($1,000,000) (0.75) = $750,000

The cost or expense to recognize would be the actual amount incurred to date, or $600,000. Therefore, the gross income that would be recognized would be calculated as follows:

| | |
|---|---|
| Revenue | $750,000 |
| Costs or expense | 600,000 |
| Gross income | 150,000 |

The use of the percentage-of-completion method is somewhat similar to the use of the cash and accrual methods in that it also recognizes revenue, expense, and income throughout the building of a project. However, it differs in that the percentage-of-completion method attempts to recognize income based on what is "earned" rather than on what is collected in cash or what is billed.

In order to contrast the accrual method and the percentage-of-completion method, let us consider the Project D data illustrated in Figure 5.2. Independent of general and administrative expenses, a gross income for Project D using the accrual method was calculated as $150,000 in Figure 5.3. The calculation is repeated in Figure 5.4.

The gross income using the percentage-of-completion method of recognizing revenue and expense is also illustrated in Figure 5.4. The difference between the two incomes in Figure 5.4 is because the construction firm has "overbilled" for work it has done. Based on calculation of costs incurred to date versus total estimated costs, the project is 66.67 percent complete. However, the firm has billed for 75 percent of its contract amount. Based on the amount of work that is completed, the construction firm should have only billed 66.67 percent of the contract amount, or $666,667.

Using the accrual method of recognizing income, the $83,333 of overbilling appears as income. Using the percentage-of-completion method, the $83,333 of overbillings would appear on the contracting firm's balance sheet as a liability in that it is money charged to a customer that is not earned to date.

When the percentage-of-completion method is used, it is very common to have over- or underbillings calculated as part of the income determination. This is primarily because the construction firm may not bill in proportion to the percent of expenditures it has incurred to the total estimated expenditures. In addition, the fact that the contracting firm cannot totally control the timing of the bills it receives from its material suppliers and vendors also can lead to over- or underbillings.

An underebilling appears as a current asset on the contracting firm's balance sheet. It represents a receivable in that it is work performed for which the firm has a reimbursement claim from the customer.

An overbilling appears on the contracting firm's balance sheet as a current liability. It is a liability in that the firm has billed (and may have received payment) for work not yet performed. In this regard the firm "owes back" the billing to its customer.

When using the percentage-of-completion method, the accounting journal entries result in recognition of revenue, expense, and income throughout a project. To illustrate accounting journal entries for the percentage-of-completion method of accounting. Let us again consider Project D data illustrated in Figure 5.4. Let us further assume single revenue and expense transactions to date of $750,000 and $600,000, respectively, as illustrated in Figure 5.4. In reality these two transactions would each be made up of several individual transactions. However, the type of accounting entry made would be the same for each separate transaction.

The $600,000 of expenditures for building Project D would be recognized by means of the following accounting journal entry:

| Construction in progress — Project D | $600,000 | |
| Cash (or accounts payable) | | $600,000 |

Billings by the contracting firm to the project owner are debited to "Accounts receivable" and credited to "Revenue billed on construction contracts". The $750,000 billing would be recognized as follows:

| Accounts receivable | $750,000 | |
| Revenue billed on construction contracts — Project D | | $750,000 |

As the cash is received from the project owner, the accounts receivable account is reduced and cash is debited. For the identified account receivable, this would be done as follows:

| Cash | $750,000 | |
| Accounts receivable | | $750,000 |

At given points in time during construction, income would be recognized for a project when using the percentage-of-completion method. For example, this may occur monthly or yearly. For purposes of illustration, let us assume income is to be recognized for Project D at the end of the year 19XO. The year-end entry based on percentage complete of costs would be

| Construction in progress — Project D | $66,667 | |
| Income projects — Project D | | $66,667 |

The $66,667 of income recognized is calculated as shown in Figure 5.4. It represents the earned profits to date based on the percentage of expenditures to date and the total estimated profit.

When Project D is complete, both the "Construction in progress — Project D" account and the "Revenue billed on construction contracts — Project D" accounts are closed out. Assuming that income is recognized periodically, the two accounts should offset. The entry made at the completion of the project would be made as follows:

| Revenue billed on construction contracts — Project D | $1,000,000 | |
| Construction in progress — Project D | | $1,000,000 |

In effect the "Construction in progress" account that is being closed includes both expenditures and recognized profit. This is why the total of the account will equal the "Revenue billed" which should equal the total contract amount.

When a loss becomes apparent during construction of a project, proper accounting calls for the recognition of the total estimated loss. The total loss should be recognized at the time it is estimated regardless of the percentage of completion at that point in time. This is consistent with the conservative principle of accounting.

## 5.18   Completed-Contract Method

The completed-contract can be used for reporting income for projects for which inherent hazards result in the inability to make reasonably dependable estimates. "Reasonably dependable" is a subjective term and has resulted in variations when the method has been used.

In practice, using the completed contract method for financial reporting is somewhat of an admission by the construction firm of a weak job-cost accounting system. The weaker the cost system is in regard to forecasting or controlling costs, the more difficult it is to determine cost to complete and percentage complete to date.

When the completed-contract method of recognizing revenue and expense is used, the firm recognizes no revenue or expense (or resulting profit or loss) until the completion of the construction project. During construction all costs incurred by the contracting firm for performing a project are accumulated in a construction in progress account which is an asset account.

Any billings to a customer by the construction firm during construction are recognized as a liability and an account entitled "revenue billed on construction contracts" is credited. In effect the revenue billed on construction contracts is a liability account until the project is complete and revenue is recognized.

In order to illustrate the accounting journal entries that would be made using the completed-contract method, let us consider the construction Project D data illustrated in Figure 5.2. While the expenditures to date would likely be made up of several labor and material expenditures, for purposes of illustrating the journal entry let us assume the expenditure is a single transaction. In reality each of the project expenditures would be made the same. Using the completed-contract method, the accounting journal entry for the $600,000 Project D expenditure would be made as follows:

| | | |
|---|---|---|
| Construction in progress — Project D | $600,000 | |
| Cash (or accounts payable) | | $600,000 |

Billings by the contracting firm to the project owner are debited to accounts receivable (an asset account) and credited to revenue billed on construction contracts. Assuming the entire $750,000 billing on Project D was made through a single billing (as is the case with expenses, this would normally be several transactions), the accounting journal entry would be made as follows:

| | | |
|---|---|---|
| Accounts receivable | $750,000 | |
| Revenue billed on construction | | |
| contracts — Project D | | $750,000 |

As the cash is received from the project owner, the accounts receivable account is reduced and cash is debited. For Project D this would be done by the following entry:

| | | |
|---|---|---|
| Cash | $750,000 | |
| Accounts receivable | | $750,000 |

No revenue or expense results from the illustrated journal entries. Only when the project is complete would revenue and expense and the resulting profit or loss be recognize. This might occur in the following year or even later. When in fact Project D is completed, an accounting entry would be made that closed out the asset and liability "holding" accounts and recognized the profit or loss. In regard to Project D the entry would be

| | | |
|---|---|---|
| Revenue billed on construction | | |
| contracts — Project D | $1,000,000 | |
| Construction in progress — Project D | | $900,000 |
| Income on projects — Project D | | 100,000 |

The effect of the entry would be to reflect $100,000 of income from Project D at the point in time when it is complete.

Whereas no income (profit) is recognized until a project is complete when using the completed-contract method, a loss must be reported as soon as it becomes apparent that a loss is to be incurred. This is consistent with the conservative principle of accounting. In fact, the entire expected loss must be recognized, not just the loss incurred to date.

A significant question that occurs when using the completed-contract method relates to when a project is to be considered complete. The date can be very important in that a matter of a few days difference can result in recognition or no recognition of significant revenues and expenses on a financial statement. A project completed on December 28 would be recognized for a December 31 year end. However, If the project is judged "completed" a mere five days later, no recognition on the statements would result.

Audit guidelines suggest that a contracting firm should be consistent in its determination of project completion dates. It also suggests that a project is complete when it is substantially complete. In the construction industry the term substantially complete usually means that the project is to the point at which it can be used for its intended purposes.

Some individuals would argue that the completed-contract method is a more accurate method of recognizing revenue and expense than the percentage-of-completion method in that the method requires no estimate of profits earned prior to the completion of the construction. These defenders of the completed-contract method argue that only at the time of a project's completion is it possible to determine profits in that prior recognition of profits is based on estimates of costs to complete. Given the uncertainty of the construction process, one might question the validity of any estimate of costs to complete a project.

The argument for completed-contract remains weak. While it may have support in that it is not dependent on estimates, the fact remains that the method results in a poor matching of revenue and expense and often results in misleading financial statements. Significant revenues or expenses may be omitted from the income statement during construction. Given this fact, the reader of the financial statements may come to the wrong conclusion regarding the profitability of a given project or the company as a whole.

## 5.19    Percentage-of-Completion versus Completed-Contract

The accounting difference between the percentage-of-completion and completed-contract methods may be dramatized with the differences in the resulting income statements that can be prepared using the two different methods.

Let us consider the data for Projects A, B, C, and D illustrated in Figure 5.2. The calculations of net income at the end of the year using both the percentage-of-completion and completed-contract methods are shown in Figure 5.5. Note that the total, recognized net income for the percentage-of-completion method is $20,000, whereas using the completed-contract method results in $180,000 loss.

Naturally the numbers illustrated in Figure 5.2 dictate the $200,000 difference in net income. Using different data for the four projects might result in a smaller dollar difference. However, the fact remains that the difference in net income reported using the two methods can be dramatic.

In Figure 5.5 only Project A is handled identically as to revenue and expense when using the two revenue-recognition methods. This is because the project is started and completed in the accounting period represented by the statement period.

The recognized loss for Project B is also handled similarly using either accounting methods. This is because the project has an estimated loss and proper accounting requires recognition of the total forecasted loss independent of the accounting method used.

Projects C and D bring out the potential dramatic difference between the percentage-of-completion method and the completed-contract method.

Neither project is completed at the time of statement presentation. Therefore, no revenue or project expense (and resulting profit or loss) is recognized in the completed-contract income statement at that point in time.
On the other hand, based on the costs expended to date and the total
estimated costs, a total of $200,000 of income is recognized on Projects
C and D. Both Projects C and D are 66.67 percent complete based on
costs expended to the total expected project costs. Therefore, two-thirds,
or 66.67 percent, of the total contract amount is recognized as revenue on
the projects.

Because of the $200,000 greater income that results when using the
percentage of completion in the year in question, it follows that in a later
year, the use of the completed-contract method will result in $200,000 of
added income. However, in the year in question, a reader of the statements
will observe the net income using one of the methods and may come to
an entirely different conclusion regarding the financial strength of the firm
than he might if he observed or analyzed the income statement using the
other revenue-recognition method.

The financial impact of the two accounting methods is not limited to
the resulting income statements. For the data illustrated in Figure 5.5, the
increased income for the percentage-of-completion method would carry over
to the firm's balance sheet as an improved working capital position relative
to that using the completed-contract method. Here, again, the reverse could
be true for a different set of project data.

Independent of the different dollar amounts possible, using either of the
two accounting methods can affect financial ratios derived from the balance
sheet as well as those derived from the income statement. The conclusion is
that the reader of the financial statements has to interpret the statements
in order to equate a statement prepared using the percentage-of-completion
method to one prepared using the completed-contract method.

Inspection of Figure 5.5 indicates that the firm's general and administrative expense (also referred to as G & A or overhead) is subtracted from
the calculated gross profit to yield the net income regardless of the choice
of accounting methods. In other words, in Figure 5.5 the G & A is written
off as a period cost (written off in the period incurred).

Some would argue that one is justified in capitalizing some of this expense in a given time period when using the completed-contract method.
The argument for this is based on the "matching" principle. Some or all of
the G & A expense shown in Figure 5.5 is the result of building Projects C
and D. Since no revenue is recognized for these two projects in the period in
question, one might defend the fact that the related G & A expense should
not be recognized until revenue is recognized. One could defend this position and therefore "defer" some or all of the $180,000 of G & A expensed on
the completed-contract net income calculation shown in Figure 5.5. Recognition of the G & A as either a period cost (expensed in period incurred)

or a product cost (capitalized as part of the asset), whereby it is different, could have been and have been defended in practice.

# 5.20  Summary

Basic to the operation of the contracting firm is the performance of financial accounting. The accounting system provides the firm the means of determining where it has been and were it is going. It is a critical part of a successful firm.

In addition to serving the needs of the contracting firm itself, the financial accounting system of the firm focuses on the needs of the many external parties that the contracting firm must satisfy. These includes the surety company, lending institutions, and public agencies. The ultimate purpose of the financial accounting process is the preparation of financial statements to include the income statement and the balance sheet.

In addition to the commonly used method of cash and accrual accounting that are used by many businesses, the percentage of completion method and the completed contract method are used in construction. The percentage of completion method recognizes that the contracting firm's interim project billings may not match the actual percent complete as determined by cost. As such, adjustments are made to the accrual method when using the percentage of completion method.

The completed contract method takes the conservative position that no income is recognized until a project is complete. In effect, it assumes that the many uncertain characteristics of the construction process prevent the determination of final project cost until the project is complete. Because of the lack of matching of income to the time period in which it incurs, the completed contract is not looked upon as a favorable method. The percentage-of-completion method is the preferred method of accounting for the contractor.

| | Project A | Project B | Project C | Project D |
|---|---|---|---|---|
| Total contract amount | $720,000 | $1,200,000 | $1,400,000 | $1,000,000 |
| Cash expenditures to 12/31/XO | 600,000 | 550,000 | 750,000 | 600,000 |
| Total expenditures to 12/30/XO | 620,000 | 700,000 | 800,000 | 600,000 |
| Estimated expenditures to complete | — | 600,000 | 400,000 | 300,000 |
| Total estimated expenditures | 620,000 | 1,300,000 | 1,200,000 | 900,000 |
| Billings through 12/31/XO | 700,000 | 800,000 | 800,000 | 750,000 |
| Cash collections through 12/31/XO | 690,000 | 700,000 | 800,000 | 700,000 |
| Cash G & A expense to 12/31/X0 = $150,000 | | | | |
| Total G & A expenses to 12/31/XO = $180,000 | | | | |

**Figure 5.2.** Data for four example projects.

| | Cash Method | Cash Method | Accrual Method | Accrual Method |
|---|---|---|---|---|
| Project A | | | | |
| Revenue | $690,000 | | $700,000 | |
| Expense | 600,000 | | 620,000 | |
| Income | | $90,000 | | $80,000 |
| Project B | | | | |
| Revenue | $700,000 | | $800,000 | |
| Expenses | 550,000 | | 700,000 | |
| Income | | 150,000 | | 100,000 |
| Project C | | | | |
| Revenue | $800,000 | | $800,000 | |
| Expenses | 750,000 | | 800,000 | |
| Income | | 50,000 | | 0 |
| Project D | | | | |
| Revenue | $700,000 | | $750,000 | |
| Expenses | 600,000 | | 600,000 | |
| Income | | 100,000 | | 150,000 |
| Total Project Income | | $390,000 | | $330,000 |
| Less G & A Expense | | 150,000 | | 180,000 |
| Net Income | | $240,000 | | $150,000 |

**Figure 5.3.** Cash versus accrual methods.

| Accrual Method — Project D | |
|---|---|
| Revenue billed | $750,000 |
| Expenditures incurred | 600,000 |
| Gross income — Accrual Method | $150,000 |
| Percentage Completion Method — Project D | |
| Expenditures to date | $600,000 |
| Total estimated expenditures | 900,000 |
| Percentage of expenditures to date | 66.67% |
| Contract amount | $1,000,000 |
| times percent complete | 66.67% |
| Should be billings — Earned Revenue | $666,667 |
| Expenditures to date | 600,000 |
| Gross income — Percentage Completion Method | $66,667 |
| Actual billings | $750.000 |
| Should be billings — Earned Revenue | 666,667 |
| Overbillings | $83,333 |

**Figure 5.4.** Accrual vs. percentage of completion method.

### Percentage of Completion

|                          | Project A | Project B   | Project C | Project D | Total       |
|--------------------------|-----------|-------------|-----------|-----------|-------------|
| Total revenue            | $720,000  | $646,155    | $933,333  | $666,667  | $2,966,155  |
| Total expenditures       | 620,000   | 700,000     | 800,000   | 600,000   | 2,720,000   |
| Additional estimated loss|           | (46,155)    |           |           | (46,155)    |
| Gross profit             | $100,000  | ($100,000)  | $133,333  | $66,667   | $200,000    |
| G & A expense            |           |             |           |           | 180,000     |
| Net income               |           |             |           |           | $20,000     |

### Completed Contract

|                     | Project A | Project B   | Project C | Project D | Total       |
|---------------------|-----------|-------------|-----------|-----------|-------------|
| Total revenue       | $720,000  |             |           |           | $720,000    |
| Total expenditures  | 620,000   |             |           |           | 620,000     |
| Forecasted loss     |           | (100,000)   |           |           | (100,000)   |
| Gross profit        | $100,000  | ($100,000)  |           |           | 0           |
| G & A expense       |           |             |           |           | 180,000     |
| Net income (loss)   |           |             |           |           | ($180,000)  |

**Figure 5.5.** Percentage of completion vs. completed contract.

# Chapter 6

# Preparation of Financial Statements

## 6.1  Introduction

The prior chapter illustrated the accounting entries for typical financial transactions of the contracting firm. These entries represent a large majority of the entries made by the contracting firm. While infrequent in occurrence, the contracting firm also has to make accounting entries for adjusting journal entries, reversing entries, and various types of changes. Examples of changes include a construction firm's switching from the completed-contract method of accounting to the percentage-of-completion method, or a change in estimate such as the estimated life of a depreciable asset.

The above accounting entries to include the adjusting journal entries and closing entries are steps required to produce the contractor's financial statements. The financial statements include the firm's balance sheet and income statement. These financial statements are needed by the firm's external firms to include the surety company and lending institutions. It is important that the contracting firm understand the preparation of the financial statements and how to interpret them.

## 6.2  Adjusting Journal Entries

Adjusting journal entries (A.J.E.) are made to recognize unrecorded transactions or incorrect account balances at the time of preparing financial statements. The number of adjusting journal entries needed will depend on the sophistication of the contracting firm's accounting process, external events that lead to financial transactions, and the time at which the financial statements are prepared. For example, if the cutoff date for preparing year-end financial statements falls on the same date as the cutoff of the weekly payroll, no adjusting entries may be needed for accruing payroll.

The purpose of adjusting journal entries is to change the general ledger records or accounts such that they will reflect the changes and cause them to be reflected correctly on the financial statements. The procedure for securing required data for adjusting journal entries consists of a systematic analysis of the ledger accounts to determine whether their balances represent the actual amount on the particular financial statement date. In many cases the adjustmentry data is obtained at the end of an accounting period by the application of foresight and the systematic arrangement of the accounts.

Some accounts are more likely to need adjusting journal entries than others. In particular, in order to reflect their account balances correctly at statement time, the following types of accounts should always be analyzed:

1. Depreciation expense
2. Accrued expenses (payroll, interest, etc.)
3. Prepaid expenses
4. Inventory
5. Accrued income
6. Prepaid income

As an example of securing adjustment data, depreciation adjustments may be determined by analyzing subsidiary records of fixed assets. An accrued payroll (an example of an accrued expense) adjusting journal entry may be determined by analyzing payroll records and noting the date of the financial statement.

Some typical adjusting journal entries (A.J.E.s) at the year-end statement preparation for a contracting firm illustrated in Figure 6.1. The A.J.E.s represent only a few of many that would often occur when analyzing the accounts of the construction firm.

The first A.J.E. in Figure 6.1 is needed because, while payday was on Friday, January 2 of the following year, on December 31 (a Wednesday) the contracting firm is indebted to workers for three days' pay. Therefore, even though no checks were written on December 31, the firm had an accrual expense as of that date.

The second adjusting journal entry made in Figure 6.1 is an example of an A.J.E. that is needed because of an unrecorded entry. Inspection of the firm's checkbook indicates that check #595 was not recorded as an expense. The adjustment has nothing to do with an accrual. Instead, it records an entry that should have been made during the year.

During a period of time, the construction firm records a number of receivables from customers on its books. This is because few customers pay cash. Some of the receivables recorded may never be paid because the customer falls upon bad financial times or has a legitimate dispute with

1. Labor expense                              7,946.00
      Accrued salaries and wages                              7,946.00
      (to show wages paid on 1-2-X1 as accrued wages on 12-31-X0)
2. Miscellaneous expense                      800.00
      Cash                                                    800.00
      (to record check #595 written by Ed for sets of plans)
3. Bad debt expense                           1,297.67
      Accounts receivable                                     1,297.67
      (to write off bad debts per Ed at 12-31-X0)
4. Accrued interest receivable                421.97
      Interest income                                         421.97
      (to record accrued interest receivable)
5. Material costs                             6,383.00
      Material inventory                                      6,383.00
      (to adjust material inventory to physical count at 12-31-X0)
6. Depreciation expense — trucks              2,922.10
      Accumulated depreciation — trucks                       2,922.10
      (to adjust accum. depreciation to actual)

**Figure 6.1.** Adjusting journal entries.

the construction firm. Therefore, it is common practice for the contracting firm to analyze its receivables at statement-preparation time and "write-off" those it deems uncollectable. An adjusting journal entry such as that shown as number 3 in Figure 6.1 is required

Adjusting journal entry number 4 in Figure 6.1 is another example of an accrual adjustment. In this case, analysis of the contracting firm's investment accounts indicates the firm has an uncollected receivable in the form of interest as of 12-31-X0. This interest would not have been recorded before the adjusting journal entry.

Because inventory is usually not a significant dollar item for many contracting firms, many choose to implement a physical inventory accounting process rather than a perpetual system. While the perpetual system is better for control purposes, the physical system is usually less time consuming as to everyday operations. Given the physical inventory process, a count of material is made at statement-preparation time. An adjusting journal entry is made to record the actual inventory. An example of such an entry is shown as number 5 in Figure 6.1. The result of the entry is to adjust the inventory shown on the construction firm's balance sheet and to recognize the appropriate material expense.

The last adjusting entry shown in Figure 6.1, number 6, is the very common entry made for acknowledging depreciation. The entry is needed because few contracting firms make the entry periodically at times be-

tween preparation of financial statements. Instead, a "catch-up" depreciation entry is made at the time of preparing financial statements. The dollar
amount of the depreciation A.J.E. recognizes the book value of the asset,
its remaining life, and the deprcciation method used.

Numerous other examples of adjusting journal entries could be given.
Because the process of detecting adjustment data for A.J.E.s is less than
scientific, it follows that it is possible to overlook the recording of all necessary A.J.E.s When this occurs, the financial statements that follow are
not totally correct. When a CPA audits the books of a contracting firm, in
effect the CPA is saying to external parties as well as the contracting firm
itself, that all material adjusting journal entries have been made. Once all
adjusting journal entries are recorded, they are summarized in the adjustments column of the working trial balance.

## 6.3   Closing the Books

Once accounting entries and adjusting journal entries are summarized on
the working trial balance, it is possible to prepare financial statements.
However, accounting entries do not terminate with the preparation of financial statements. After the statements have been prepared, the ledgers
are closed which is commonly referred to as the closing of the books. This
means that the balances of all the adjusted revenue and expense accounts
are transferred to the income summary account. In effect the income summary account represents the profit or loss for the contracting firm for the
period in question. This amount is transferred to the retained earnings of
the corporation or net worth of the proprietorship. The described process
of closing the books is illustrated in Figure 6.2. The contracting firm would
have more revenue and expense accounts to close than those shown in the
figure; however the processes would be the same.

The result of the process illustrated in Figure 6.2 is to reset the balances
of all revenue and expense accounts to zero. This is necessary in order to
divide the financial accomplishments of the construction firm into periods.

## 6.4   Reversing Entries

Once all entries have been made for preparing financial statements and
closing entries have been made to initiate the start of a new reporting
period, it is often convenient to reverse certain balance sheet accounts that
serve as temporary accounts.

Let us consider adjusting journal entry number one illustrated in Figure 6.1. The entry was made at the time of preparing the financial statements. The entry was as follows:

| Revenue — contract sales | $850,000 | Labor expense | $300,000 |
|---|---|---|---|
| Interest income | 20,000 | Material expense | 300,000 |
| | | Equipment expense | 100,000 |
| | | Office salaries | 50,000 |
| | | Depreciation | 70,000 |

Closing entry #1 (to close revenue and expense accounts)

| Revenue — contract sales | $850,000 | |
|---|---|---|
| Interest income | 20,000 | |
| Labor expense | | $300,000 |
| Material expense | | 300,000 |
| Office salaries | | 100,000 |
| Depreciation | | 50,000 |
| Income Summary | | 70,000 |
| | | 50,000 |

Closing entry #2 (to transfer profit to retained earnings on balance sheet)

| Income summary | $50,000 | |
|---|---|---|
| Retained earnings | | $50,000 |

**Figure 6.2.** Closing entries.

| Labor expense | $7,946 | |
|---|---|---|
| Accrued salaries and wages | | $7,946 |

The effect of the entry is to record an expense for the contracting firm for the year-end 20X0. In addition, a balance sheet liability — accrued salaries & wages — would be created.

When the contracting firm closes its books, it will close out the labor expense account balance by crediting the labor expense account and debiting the income summary account. This process was illustrated in Figure 6.2.

Now let us assume that we are into the following year, year 20X1. The contracting firm on 1-2-X1 pays its payroll. Let us assume the week's payroll is for $15,000. It might make the following entry:

| Labor expense | $15,000 | |
|---|---|---|
| Cash | | $15,000 |

However, note that, of the $15,000 expense recognized, in reality $7,946 of it was for an expense incurred in prior year and recognized as an expense in the financial statements for the year end 20X0. In effect, the firm has recognized $7,946 twice. In addition, note that using the procedure described, the balance sheet item — Accrued salaries and wages $7,946 — would maintain its balance throughout the 20X1 accounting period.

Naturally the firm might remember that part of the newly created labor expense is for an expense of the prior period and therefore only recognize the difference, $7,054, as a new expense. However, given many of these types of transactions, the contracting firm may fall victim to double recognition of entries.

To avoid the possible doubling up of the expense recognized in prior year and to remove the temporary balance sheet liability account item, the following reversing entry is made after the books are closed.

| Accrued salaries and wages | $7,946 | |
|---|---|---|
| Labor expense | | $7,946 |

This reversing entry removes the balance sheet temporary account and creates a credit labor expense balance for the new year, 20X1. This credit balance will offset a portion of the $15,000 expense when paid in 20X1. The result will be that the recognized labor expense will be as follows:

| 20X1 | $15,000 |
|---|---|
| Less reversing entry | 7,946 |
| Labor expense incurred in 20X1 | 7,054 |

By reversing the adjusting journal entry number 1 in Figure 6.1, it is not necessary for the construction firm in 20X1 to remember what the A.J.E.s were at year-end 20X0. It merely records the 20X1 entry as it would any other entry that occurs in 20X1.

The reversing of entries described is a matter of convenience. The process does away with the necessity of analyzing expense payments made in the new period to determine what portion of each payment belonged to the past period and what part should be charged to the present period.

Usually all balances of accrued accounts and prepaid accounts are reversed including both revenue and expense accounts. However, any reversing entry is for convenience and is not a mandatory account entry. The reversing entry is intended to facilitate subsequent accounting.

## 6.5    Accounting Changes

Regardless of the sophistication of the construction firm's accounting process, situations frequently occur in which changes in accounting methods, estimates, or error corrections must be made. Unlike the routine accounting entries described in the previous chapter, which are handled rather uniformly in practice from one firm to the next, practice has seen a somewhat diverse range of accounting treatment for various types of accounting changes. Only recently, since the accounting profession has set forth several opinions on handling changes, has accepted treatment of changes been defined.

An accounting change can be classified as one of the following three types:

1. Change in accounting principle
2. Change in accounting estimate

3. Change due to discovery of an error

Each of these types of changes is handled somewhat differently as to required journal entries and resulting financial statements.

A change in accounting principle occurs when the contracting firm adopts a different, generally accepted accounting principle or procedure from one previously used. An example would be a change from using the straight-line depreciation method to the double-declining depreciation method. Another example would be a change from using the FIFO (First In First Out) method of valuing inventory to the use of LIFO (Last In First Out) for valuing inventory.

A change in estimate occurs when the contracting firm becomes aware of more accurate determinations of fact than what is initially estimated in a prior period. An example would be when the construction firm learns that the life of a piece of equipment is eight years rather than the initial estimate of six years. The revised estimate may be made after a few years of using the equipment.

A change due to an error discovery is defined as a change necessitated by the misapplication of facts existing at the time of the initial accounting entry. The misapplication may have been because of oversight or it may have been intentional. An example of a misapplication of facts would be expensing an item of construction equipment in a prior period when the equipment should have been capitalized in that it had an estimated life of three years.

The significant issue about treatment of accounting changes is the impact the treatment has on the financial statements of the contracting firm. In particular, the need to recognize an accounting change raises the question of whether it should be handled on the financial statements in the current accounting period, retroactively to prior accounting periods, or prospectively over future accounting periods. The acceptable treatment for the identified types of accounting changes is discussed separately for each type of change in the following sections.

## 6.6  Financial Statements

The entire financial accounting process may be viewed as being aimed at preparing financial statements. While financial statements may be prepared to meet the needs of external parties, the statements can also serve an internal purpose for the contracting firm in making both short-term and long-term decisions. Decisions such as determining whether equipment should be purchased or leased or what type of financing should be obtained for operations or capital expenditures are aided by prepared financial statements.

In practice the frequency of the contracting firm's preparation of financial statements varies. As a minimum, financial statements may be prepared annually enabling income tax return preparation and summarizing the year's results.

To satisfy external or internal needs, financial statements may be prepared more frequently than yearly. Quarterly or monthly statements are becoming more common in the industry.

## 6.7   Types of Financial Statements

The financial statements of a firm may be limited to a statement of financial position and an income statement. These are by far the most common financial statements prepared in practice. The statement of financial position is more commonly referred to as a balance sheet. The income statement is also commonly referred to as the profit and loss statement (the P & L statement) or the statement of operations.

Current accounting guidelines indicate that the correct terminology for the two mentioned statements are the statement of financial position and the income statement. However, the use of balance sheet remains common in practice as an alternative to calling the statement, the statement of financial position. In recognition of this, we will use the term balance sheet, Use of the term *income statement* has gained in popularity versus use of the term *profit and loss statement*. Therefore, we will use the term *income statement* or *statement of operations*.

The balance sheet reports on the assets, liabilities, and net worth, or owner's equity accounts, at a given point in time. On the other hand, the income statement refers to a period of time: it reports on the revenues and expenses for that period. Because the balance sheet presents a point in time, it can be viewed as a snapshot of the firm's financial status; whereas the income statement, which represents a period of time, can be viewed as a film of the financial operation.

If a contracting firm desires audited financial statements prepared by a CPA, the AICPA requires that a third statement be prepared in addition to the balance sheet and income statement. This is referred to as the statement of changes in financial position or, less appropriately, as the funds statement.

In addition to the three financial statements noted, audited financial statements will also include an opinion statement of the auditor, disclosure of significant accounting policies of the audited construction firm, and notes that further describe line items presented on the financial statements.

The contracting firm is typically dependent on the financial statement needs of external parties. These external parties include surety (bonding) companies, lending institutions, government agencies, material vendors, and equipment financing companies. Some of these external parties will require financial information or reports in addition to those required to meet the minimum standards of properly prepared, audited financial statements. For example, a statement of work in progress is often requested from the contracting firm.

In addition to requested financial reports, financial statements are often accompanied by various schedules or statements that support item totals of the previously noted three required financial statements. These schedules or statements usually consist of a statement of general and administrative expenses (also referred to as operating expenses) and a statement of retained earnings. Less frequent in preparation are schedules or statements on amounts due on contract and statements of miscellaneous income such as rental property revenue and expense schedules.

## 6.8   The Working Trial Balance

Preparation of financial statements including the balance sheet and income statement is initiated with the preparation of a working trial balance which is also referred to as the WTB or sometimes as the worksheet.

The WTB is not a financial statement in itself. Instead, it serves as a worksheet for preparing financial statements. The worksheet provides a means of summarizing the account balance of each of the firm's ledger accounts at the time of financial statement preparation. Each and every account is listed regardless of the account balance. An example of a WTB format for a construction firm is shown in Figure 6.3. Only representative accounts and the debit and credit entries for these accounts are shown.

Following the listing of the account titles and code numbers in the WTB is the balance of the account as adjusted for the end of the previous year. This is followed by the balance of each account as per the general ledger books at the end of the current year.

The balances are shown in the debit and credit columns depending upon whether the specific account has a debit or credit balance. Assuming the firm's set of books is in balance, the debit and credit totals in the WTB should equal one another.

In order to recognize accrual and overlooked accounting entries at the time of preparing financial statements, various adjusting journal entries (A.J.E.s) must be made to the firm's set of books. Examples of A.J.E.s for a WTB were shown in Figure 6.1. These adjusting journal entries would be recorded in the appropriate columns in the WTB as illustrated in Figure 6.3. Note that not all of the A.J.E.s are illustrated in Figure 6.3.

The adjusting journal entries listed in the WTB are added to the account year-end balances per book to yield the adjusted account balance. These are also shown in Figure 6.3.

Finally, balance sheet account balances are separated from income statement balances by listing them in separate columns as shown in Figure 6.3. In order to equate the debit and credit columns in both the balance sheet and income statement, a "plug" number, representing the income or loss for the period, must be determined.

Preparation of the balance sheet and income statement comes directly from the account balances shown on the prepared WTB. Typically, financial statements do not itemize every chart of account dollar balance; instead, several account balances shown on the WTB will be summed and reflected as a single line item.

## 6.9   Balance Sheet

The balance sheet is prepared directly from the working trial balance illustrated in Figure 6.3. The balance sheet accounts listed in the WTB are grouped and presented as line items on the balance sheet at a specific point in time, such as the firm's year end.

The balance sheet, or statement of financial position, reports on the basic economic equation of the firm:

Assets = Creditor's Equity (Liabilities) + Owner's Equity

While individual types of accounts may vary from one firm to the next, the basic economic equation holds true for every firm. For example, the types of owner's equity accounts differ for the individual proprietor and the corporation. However, each type of firm has equity accounts that have the effect of balancing the asset and creditor's equity portion of the economic equation.

The creditor's equity segment of the basic economic equation for a firm is more commonly referred to as the liabilities of the firm. When the firm is a corporation the owner's equity segment is commonly referred to as stockholder's equity.

The broad-based economic equation is further subdivided into classifications that provide the means of analyzing, interpreting, and comparing the financial statements with past statements and statements of other firms. Although the classifications vary somewhat from one firm to the next depending on the type of firm, the following classifications are typical:

| Assets: | Current assets |
|---|---|
| | Investments and funds |
| | Fixed assets — tangible |
| |    Fixed assets — intangible |
| | Other assets |
| | Deferred charges |
| Liabilities: | Current liabilities |
| | Long-term liabilities |
| Owner's equity: | Contributed capital |
| |    Capital stock |
| |    Contributions in excess of par |
| | Retained earnings |
| |    Appropriated |
| |    Unappropriated |

Most of the account classifications are self-explanatory. For example, current assets are those that are readily available and likely to be used in performing the firm's operations in the current account period. Other assets are those assets that do not fit the other asset classifications. In particular, a receivable from one of the firm's officers and a company office that is currently not being used in the firm's production process fit into the other asset category.

The difference between appropriated retained earnings and unappropriated retained earnings lies in whether the retained earnings are restricted or not restricted as to being available for payment of dividends or other uses. In particular, some retained earnings may be restricted in the firm's agreement with creditors. The creditors may restrict the payment of dividends in order to assure the creditors that retained earnings are available for bond interest and principle payments.

A typical contracting firm's balance sheet position is shown in Figure 6.4. As can be observed from the statement, several accounts unique to the construction industry are a part of the statement. For example, "contract equity not billed" is a current asset account somewhat unique to the construction industry. In effect, it is a type of receivable. As previously discussed, when the costs of projects in progress exceed the billings, the difference is shown as a current asset. It should be noted that a current liability is required to be shown when billings are in excess of costs. No overbillings are assumed in the balance sheet illustrated. When under- and overbillings both exist, they should be shown separately as an asset an a liability on the balance sheet. They should not be netted. The actual calculation of the amounts would probably be shown by means of a note to the financial statement or a supporting schedule. A schedule may also be prepared to support the amount listed for plant, property, and equipment.

While the amounts shown in the balance sheet in Figure 6.4 are for a given point in time, comparative financial position statements are sometimes prepared. This is done by showing the financial status of each account listed at the current point in time and at some past point in time such as one year prior. Such statements aid the reader in analysis of the statement.

## 6.10   Income Statement

The income statement is merely the summation of the firm's revenues and expenses for a given period of time. The dollar amounts on the statements also come from a WTB such as prepared in Figure 6.3. The relevant period of time is normally a quarter of a year or an entire year. However, when in need of a bank loan, a firm may be required to prepare an income statement for a period of time as short as a month.

The income statement has grown in significance over the past few years. Whereas creditors and investors of prior years placed their emphasis on the analysis of the firm's financial position statement, the income statement has played an increasingly significant role. Today's investors are primarily concerned with a firm's ability to continually increase its reported annual operating and net income.

As discussed in an earlier section, and as is true in regard to the balance sheet, the form and individual accounts included in the income statement vary in practice. Some statements may report all revenue (operating and non-operating) first followed by a listing of all expenses, whereas other statements may first list operating revenues and expenses followed by non-operating revenues and expenses. In addition, some statements may provide a combined income statement and retained earnings statement, whereas in other case the income statement may be presented independently of the retained earnings statement. The income statement purpose remains the same regardless of the form of the statement which is to yield the net income for the firm for a given period of time. In addition, if the firm is a corporation, a calculation yielding the net income earned per share of company stock (indicative of the firm's rate of return on investment) is made and the earnings per share (EPS) is reported on the income statement.

Typical construction industry accounts summarized in an income statement (also referred to as a Statement of Operations) are shown in Figure 6.5. As discussed in an earlier section, the accounts and dollar amounts to be reported in the income statement differ depending on whether the percenage-of-completion or completed-contract method is used for recognizing revenue. The statement shown reflects using the percentage-of-completion method. The income statement shown in Figure 6.5 is for the same firm represented in the balance sheet in Figure 6.4. Note the tie or relation between the income dollar amount and the change in retained earnings on the balance sheet.

Operating income is often separated from extraordinary income on the income statement. Extraordinary income is significantly different in character from the typical business activity income of the firm. It has a material effect on income and its occurrence is not expected to recur frequently. A contracting firm's selling part of its central office, or condemnation of part of its branch offices, would be considered extraordnary items on the firm's income statement. Separation of operating income and extraordinary income is important in that a better analysis of the firm can be made. In particular, most of the attention is drawn toward the operating income trend in that extraordinary items, while strongly influencing net income, are viewed as occurring only once.

# 6.11   Statement of Changes in Financial Position

Not as widely published or recognized as the financial position or income statement, but of increasing significance and frequency, is the statement of changes in financial position. This statement is viewed as essential to the fair reporting of the causes of the changes in the financial position (i.e., the balance sheet) from one period to the next. While the income statement reports changes as a result of operations, it does not reflect all of the financial changes between two consecutive accounting periods. As such, the statement of changes in financial position has grown in acceptance and importance to potential investors and creditors.

The statement of changes in financial position can be prepared on a working capital basis or on a cash basis. Regardless of which basis is used, the preparation of the statement of changes in financial position is less straightforward than preparation of the income statement or the financial position statement. Whereas the income statement and statement of financial position are merely the representation of summed accounts, the statement of changes in financial position requires the analysis of transactions and has three specific purposes:

1. To report on all the financing and investing transactions of the firm.
2. To report on the generation and application of funds (either on a working capital or cash basis).
3. To report the causes of all the changes in financial position during the period.

# 6.12   Miscellaneous Schedules

The reader of the construction firm's financial statements may require additional information that is not disclosed on the financial statements. This often takes the form of additional financial statements referred to as schedules.

The schedules support various line items that are presented on the balance sheet, income statement, and statement of changes in financial position. The schedules supply information to banks and sureties that enhance their ability to analyze the contracting firm. The schedules also assist the owners or manager analyze the company's financial position and results of operations which can lead to necessary decisions. Financial statements that include several supporting schedules are sometimes referred to as long-form financial statements.

The two most common additional financial statements are the schedule of indirect cost and operating expense and the schedule of contracts in progress. Other somewhat common schedules or statements are the retained earning statement and schedule of notes payable.

An example of a schedule of indirect cost and operating expense is illustrated in Figure 6.6. The schedule is for the same firm illustrated in the other financial statements in this chapter. The sum of the indirect costs and operating expenses shown in Figure 6.6 is also shown in the statment of operations illustrated in Figure 6.5. In effect, the schedule of indirect cost and operating expense "explains" the total dollar amount shown in Figure 6.5 by itemizing the component costs that make up the total. This itemization provides the reader a means of analyzing what may be a significant expense item on the income statement or statement of operations. For example, by reviewing the schedule, the reader can differentiate between ordinary costs of doing business and what might be considered "perks" or unusual types of expenditures.

The schedule of contracts in process, or work-in-process schedule, is a supporting financial statement somewhat unique to the construction firm. Its purpose is to clarify various items on the income statement and to establish the workload at the financial statement date.

An example of a schedule of contracts in process is shown in Figure 6.7. Note that several of the dollar totals on the schedule of contracts in process are shown on the balance sheet shown in Figure 6.4. For example, the sum of the "contract equity not billed" is shown on the balance sheet as a line item.

Because of the relatively long-term nature of the construction project and the relatively large dollar value of the work it performs, it is common for the contracting firm to have a significant dollar amount of work in process at any one point in time. Depending on the method of revenue recognition the firm uses, the revenue and related expense of work in process would be reflected differently on the financial statments. The schedule of contracts in process provides the reader a means of interpreting the statements. For example, from this schedule the reader can evaluate the following:

- Profit margins for various types of projects
- Backlog of work in volume or cost of work to complete
- Cash to be generated
- Over or under billings

Perhaps the greatest need or justification for the schedule of contracts in process relates to the question of the profitability of the work in process. When the dollar amount of work in process is substantial, the potential profit (or loss) of the work can have a large impact on the firm's ability to continue as a going concern. By preparing a schedule such as the one illustrated in Figure 6.7, the statement reader is in a better position to evaluate the financial strength of the company.

## 6.13   Summary

Financial statements are the means of communicating the financial results and financial well being of the contracting company. The primary financial statements are the balance sheet and the income statements.

These statements are the summary of the daily financial transactions that happen every day in the firm. These transactions are entered into journals, posted into ledgers, and summarized via a working trial balance. As part of the process of preparing the financial statements, adjusting journal entries are made, ledgers are reversed, and closing entries are made.

In addition to the financial statements, the unique characteristics of the construction firm result in the need to produce supporting schedules. Especially important is the work in process schedule. This schedule reports on the individual status of projects in progress at the time of preparing the financial statements.

| Account[a] | Trial balance 12/31 prior year DR | CR | Trial balance 12/31/X0 DR | CR | Adjustments DR | CR | Balance sheet DR | CR | Income statement DR | CR |
|---|---|---|---|---|---|---|---|---|---|---|
| Petty cash | 250 | | 991 | | | 800 | 191 | | | |
| Cash in bank | 5,480 | | 4,550 | | | | 4,550 | | | |
| Savings | 38,640 | | 40,400 | | | | 40,400 | | | |
| Accounts receivable | 298,400 | | 318,270 | | | | 318,279 | | | |
| Material | | | 31,861 | | | 6,383 | 25,478 | | | |
| Accounts payable | | 58,219 | | 57,163 | 7,946 | 65,109 | | | | |
| Accrued workmen's comp. | | 11,205 | | 13,058 | | | | 13,058 | | |
| Notes payable | | 93,200 | | 81,950 | | | | 81,950 | | |
| Common stock | | 95,000 | | 95,000 | | | | 95,000 | | |
| Capital contributions | | 60,000 | | 60,000 | | | | 60,000 | | |
| Retained earnings | | 231,400 | | 295,335 | | | | 295,335 | | |
| Contract sales | | 2,300,200 | | 2,500,500 | | | | | | 2,500,500 |
| Contract work in process | | 190,400 | | 100,500 | | | | | | 100,500 |
| Time and material sales | | 42,650 | | 70,360 | | | | | | 70,360 |
| Labor cost | 638,566 | | 633,231 | | 7,946 | | | | 641,177 | |
| Material cost | 843,780 | | 886,221 | | 6,383 | | | | 892,604 | |
| Subcontractors | 410,650 | | 360,896 | | | | | | 360,896 | |
| Other direct cost | | | 11,550 | | | 800 | | | 10,750 | |
| Officer's medical exp. | 1,750 | | 1,885 | | | | | | 1,885 | |
| Provision for taxes | 21,950 | | 23,800 | | | | | | 23,800 | |
| | 3,510,443 | 3,510,443 | 3,560,772 | 3,560,772 | 18,293 | 18,293 | 889,412 | 889,412 | 2,615,210 | 2,671,360 |
| | | | | | | | | | 56,150 | |
| | | | | | | | | | 2,671,360 | |

**Figure 6.3.** Example working trial balance.

[a]Note: Columns don't foot because only representative accounts are shown. All accounts would be shown in a working trial balance and column totals would equal those shown.

| ASSETS | | LIABILITIES | |
| --- | --- | --- | --- |
| **Current Assets** | | **Current Liabilities** | |
| Cash on hand and in bank | $45,141 | Accounts payable: | |
| Accounts receivable: | | Trade | $48,504 |
| Customers | 198,224 | Erectors Inc | 2,111 |
| Fabricators Inc | 7,040 | Fabricators Co | 65 |
| Steel Erectors Inc | 3,860 | Trucking Co Inc | 12,105 |
| Redi-Mix Co | 179 | Retained percentage payable | 2,324 |
| Investment Co | 3,791 | Equipment contracts payable | 20,352 |
| Officers and employees | 10,733 | Bank notes payable: | |
| Retained percentages receivable | 94,452 | Unsecured | 7,200 |
| Notes receivable: | | Secured | 50,203 |
| Fabrication Co Inc | 50,000 | Notes payable — unsecured: | |
| Officer | 16,000 | Officers | 41,950 |
| Accrued interest receivable | 572 | Others | 40,000 |
| Time and material not billed | | Withheld and accrued payroll taxes | 13,058 |

| | | |
|---|---:|---:|
| Customers | 4,383 | |
| Investment Co | 571 | |
| Officers and employees | 2,459 | |
| Contract equity not billed | 28,746 | |
| Inventory of material and supplies | 25,478 | |
| Prepaid insurance | 2,236 | |
| Prepaid rent | 2,000 | $495,865 |
| **Investments** | | |
| Cash value of life insurance | 41,279 | |
| Pleasure driveway—district bonds | 6,000 | |
| Lane bonds—at cost | 3,000 | 50,279 |
| **Property and Equipment** | | |
| Machinery and equipment | $437,585 | |
| Office furniture and equipment | 42,270 | |
| Autos and trucks | 157,597 | |
| Total—at cost | 637,452 | |
| Less cost charged to operations | 294,184 | 343,268 |
| **TOTAL ASSETS** | | $889,412 |

| | | |
|---|---:|---:|
| Other accrued expense: | | |
| Salaries and wages | 70,269 | |
| Interest | 3,925 | |
| Property taxes | 982 | |
| Estimated liability for income taxes | 13,131 | $326,179 |
| **Long-Term Liabilities** | | |
| Equipment contracts payable | 69,615 | |
| Notes payalbe to insurance companies | 21,283 | |
| Other notes payable—unsecured | 22,000 | 112,898 |
| **Commitments and Contingencies** | | |
| **Stockholders' equity** | | |
| Common stock—950 shares | 95,000 | |
| Capital contributed in excess of par | 60,000 | |
| Retained earnings | 295,335 | 450,335 |
| **TOTAL LIABILITIES & EQUITY** | | $889,412 |

**Figure 6.4.** Balance sheet.

| | | |
|---|---:|---:|
| **Revenue** | | $2,671,360 |
| **Cost of Sales** | | |
| Material | $892,604 | |
| Subcontracts | 360,896 | |
| Direct labor | 641,177 | |
| Equipment rental | 20,853 | |
| Other direct cost | 10,750 | 1,926,280 |
| Gross profit on sales | | $745,080 |
| **Indirect cost and operating expense** | | 692,023 |
| Net income on operations | | 53,057 |
| **Other income** | | |
| Interest | 2,316 | |
| Discounts earned | 15,987 | |
| Gain on sale of equipment | 890 | 19,193 |
| Net income before income taxes | | $72,250 |
| **Income taxes** | | |
| Provision for income taxes | $23,800 | |
| Investment credit | 7,700 | 16,100 |
| | | $56,150 |
| **Net income for the year** | | |
| Earnings per share, $59 | | |
| **Retained earnings** | | |
| Balance, January 1, 20X0 | | $239,185 |
| Balance, December 31, 20X0 | | $295,335 |

**Figure 6.5.** Statement of operations and retained earnings.

| | |
|---|---:|
| Salaries and wages | $151,219 |
| Payroll taxes | 53,227 |
| Union welfare and benefits expense | 55,896 |
| Travel and entertainment | 9,462 |
| Auto and truck expense | 74,704 |
| Tools and supplies | 23,806 |
| Repairs and maintenance | 40,614 |
| Depreciation | 96,051 |
| Rent | 9,750 |
| Officers medical expense | 1,885 |
| General insurance | 22,437 |
| Workmens compensation insurance | 24,455 |
| Officers life insurance | 4,036 |
| Office supplies and postage | 5,746 |
| Pension plan expense | 24,795 |
| Legal and accounting | 12,239 |
| Advertising | 2,383 |
| Dues and subscriptions | 1,745 |
| Contributions | 1,879 |
| Property and other taxes | 2,581 |
| Utilities | 7,489 |
| Interest | 38,085 |
| Bad debt | 17,704 |
| Miscellaneous | 9,235 |
| Management fees | 600 |
| Total | $692,023 |

**Figure 6.6.** Schedule of indirect cost and operating expense.

| | Contract price | Contract inc. earned to date | Contract billing to date | Contract equity not billed | Total estimated cost | Contract cost incurred to date | Estimated cost to complete | Estimated total gross profit |
|---|---|---|---|---|---|---|---|---|
| Lutheran Church | $68,187 | $19,000 | $19,000 | | $42,700 | $11,913 | $30,787 | $25,487 |
| Storage System | 265,139 | 253,097 | 253,097 | | 218,460 | 208,411 | 10,049 | 46,679 |
| North Street | 111,857 | 106,614 | 106,614 | | 90,318 | 85,983 | 4,335 | 21,539 |
| Tire Co | 160,476 | 146,311 | 146,311 | | 139,896 | 127,585 | 12,311 | 20,580 |
| Estates Co | 252,721 | 121,575 | 121,575 | | 237,222 | 114,104 | 123,118 | 15,499 |
| South Avenue | 42,966 | 31,162 | 31,162 | | 39,771 | 28,516 | 11,255 | 3,195 |
| High School | 111,343 | 89,215 | 89,215 | | 95,235 | 75,521 | 19,714 | 16,108 |
| Boat Club | 17,545 | 17,000 | 17,000 | | 11,882 | 11,514 | 368 | 5,663 |
| Summit | 21,183 | 19,574 | 19,574 | | 19,340 | 17,870 | 1,470 | 1,843 |
| East Avenue | 382,774 | 53,206 | 24,460 | 28,746 | 344,500 | 47,879 | 296,621 | 38,274 |
| Parking Deck | 108,336 | 21,016 | 21,016 | | 60,935 | 11,821 | 49,114 | 47,401 |
| State Bank | 31,728 | 28,183 | 28,183 | | 23,020 | 22,053 | 967 | 8,708 |
| Total | $1,574,255 | $905,953 | $877,207 | $28,746 | $1,323,279 | $763,170 | $560,109 | $250,976 |

**Figure 6.7. Work in progress schedule.**

# Chapter 7

# Financial Management

## 7.1 Introduction

Numerous reasons are given for the high rate of bankruptcy of contractors in the construction industry. Blame is often aimed at the potentially inaccurate estimating and lack of project controls that characterize the construction firm. Too often one fails to recognize a construction firm's inherent financial structure shortcomings. A firm's financial statements, including the balance sheet and income statement, are the means to evaluate its financial structure and analyze its shortcomings.

External parties such as sureties and lending institutions must review and analyze the financial statements of the contracting firm in order to avoid being party to a financial failure. The contractor firm itself should objectively review and analyze its financial statements. This can lead to profitable decision making and can prevent actions or decisions that potentially lead to failure.

There are few if any types of businesses that are as dependent on external parties as the contracting firm. These external parties often rely on their analysis of the contracting firm's financial statements as their means of establishing their business dealings with the firm. In many instances, the contracting firm's ability to continue as a going concern depends on external parties' interpretation of the construction firm's financial statements. For example, a surety's willingness to write a bond for a contracting firm (a bond that may be required to enable the firm to do work) will be very dependent on the surety's appraisal of the firm's strength as outlined in the financial statements.

This chapter discusses the contracting firm's financial strength as illustrated by the financial statements. Analysis of the statements is discussed along with factors that external parties evaluate in appraising the firm's financial pluses and minuses.

## 7.2    Construction Firm Financial Structure Characteristics

Unique external factors tend to shape the total financial structure of the contracting firm. Inability to raise equity capital, operating on short-term credit, large amounts of receivable on the books because of the owner's retention of partial payments, and dependence of some firms on large dollar values of equipment all dictate the financial structure. Some of these industry characteristics result in a less than favorable financial structure that contains an undue amount of risk. While any one of these financial factors may not be representative of every firm, they tend to be characteristic of the typical contracting firm.

The primary assets of any firm are its cash, receivables, inventory, and fixed assets. Compared to all other industries, the contracting firm has a high percentage of its assets in receivables. In addition, compared to the manufacturing industry (of which construction is a part), as illustrated in Figure 7.1, the contracting firm has a relatively small percent of fixed assets.

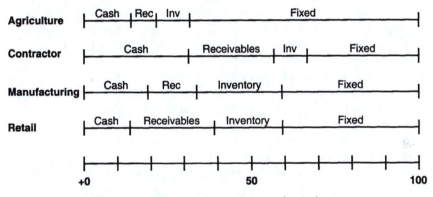

**Figure 7.1.** Comparison of assets by industry.

The contracting firm's high dependence on receivables results from the typical billing and payment retention procedures. Upon completing various work items, the contracting firm bills the owner. However, depending on the client, actual payment may be slow. Delays of a month or more are common. In addition, when in fact the payment is made, the client may choose to retain part of the payment (typically 10 percent) until the construction project is substantially completed.

Receivables are troublesome assets. For one thing they do not provide the contracting firm with an asset that can be used for operations and growth. Receivables typically do not aid the firm in resolving any cash-flow problems. In fact a large buildup of receivables is often characteristic of a cash-flow problem leading to financial difficulties.

Unlike receivables in other industries, the contracting firm seldom receives interest from the customer. The customer is in fact using the contracting firm's money to finance his project. The uncertainty and risk associated with receivables are other unfavorable financial considerations.

Various construction-related associations have continued their efforts to alleviate the difficulties that are part of the process by which the contracting firm is paid. However, by itself the individual contracting firm can do little to change the process. As such, the firm has to pay special attention to the control of its receivables which entails adequate collection procedures and an adequate and realistic provision for bad debts.

Inspection of Figure 7.1 also reveals that the contracting firm has a relatively large percentage of its assets in cash. The contracting firm continually has to have cash on hand to pay for varying amounts of labor and material expenses. While an adequate amount of cash for this purpose is necessary in order to avoid cash-flow problems, the fact remains that cash is not a good income-producing asset.

Cash-heavy firms tend to be non-growth firms in that income and growth are generated by means of investment in non-monetary assets such as plant and equipment. The contracting firm has to be cautious in using its cash for non-monetary investments because it needs to have large amounts of cash readily available for operations.

It is difficult to generalize about the contracting firm's investment in fixed assets. The amount of such assets within a given contracting firm varies from relatively high to practically nothing. However, as can be observed from Figure 7.1, the typical contracting firm has less fixed assets than its counterparts in the manufacturing industry. While this smaller dependence on fixed assets frees cash for other purposes, it is also true that firms with relatively large amounts of investment in such fixed assets tend to be more financially stable.

On the capital side of the contracting firm's balance sheet, the firm is characterized by a high degree of debt. Whereas typical manufacturing firms have approximately the same amount of debt capital and equity capital (i.e., invested capital plus retained earnings), the contracting firm typically has liabilities in excess of equity capital. This high dependence on debt is not unexpected when one considers the difficulty that the contracting firm has with raising investment capital. Typically, the contracting firm is relatively small and not in a favorable position to issue stock that provides equity capital.

Usually a large percentage of a contracting firm's liabilities derives from several factors. For one, as with investment capital, third parties viewing the contracting firm as risky are reluctant to commit funds to the firm on a long-term basis. Secondly, many contracting firms lease their construction equipment by means of a long-term contract.

The contracting firm's high dependence on current liabilities place it in a position of high financial risk. By definition, current liabilities come due in a relatively short period of time. Should the firm have difficulty making payments because of a slump in its operations or poor cash flow planning, it may default and become subject to bankruptcy proceedings.

## 7.3 Statement Analysis Difficulties

Alternative accounting methods and means of expressing financial data in financial statements are especially variable in the construction industry. Lenders and sureties have fallen victim to misleading financial statements. At the same time, the contractor's ability to continue as a going concern may depend more on the firm's accountant's ability to present financial statements in the most favorable light rather than the firm's technical knowledge or actual financial structure.

To illustrate the significant difference that the choice of accounting method can have on the construction firm's financial statements, let us consider data regarding Contracting Co. Inc., a newly started corporation as of January 1, 20XO. The owner of the company started the firm by investing $50,000 cash in the firm of which $10,000 was used to purchase material inventory and $10,000 was used to purchase 5 percent equity in $200,000 of equipment with an average life of ten years. The owner used an additional $20,000 of his personal cash to purchase a $100,000 building for an office.

Contracting Co., Inc. engaged in two projects: Project 1 was incomplete at 20XO year end while Project 2 was completed during the year. Data regarding the two projects are as follows:

| Project | Total contract price | Contract expenditures to 12/31/X0 | Estimated additional cost to complete | Cash collections to 12/31/X0 | Billings to 12/31/X0 |
|---|---|---|---|---|---|
| 1 | $280,000 | $187,500 | $12,500 | $190,000 | $190,000 |
| 2 | $350,000 | $300,000 | | $320,000 | $350,000 |

An additional $15,000 was spent for material inventory during the year. The material inventory was in inventory at 20XO year end. Overhead costs are assumed to be $10,000.

Shown below are two abbreviated income statements.

| | State-ment 1 | State-ment 1 | State-ment 2 | State-ment 2 |
|---|---|---|---|---|
| Revenue | | $612,500 | | $350,000 |
| Less cost | | −487,500 | | −300,000 |
| Depreciation (eq) | | −20,000 | | −30,000 |
| Depreciation (bldg) | | | | −3,750 |
| Beginning inventory | $10,000 | | $10,000 | |
| Purchases | 15,000 | | 15,000 | |
| Ending inventory | −7,500 | | −5,000 | |
| Material costs | | −17,500 | | −20,000 |
| Gross profit (loss) | | $87,500 | | −$3,750 |
| Overhead expense | | −10,000 | | −10,000 |
| Net inventory before tax | | $77,500 | | −$13,750 |

Statement 1 reflects a $77,500 profit while Statement 2 indicates a $13,750 loss. The interesting point is that both statements are income statements for the same firm, i.e., Contracting Co., Inc. for the year 20XO. How can it be that both income statements reflect the same firm? The answer is that different accounting methods are used in Statement 1 versus Statement 2.

Statement 1 recognizes revenue using the percentage-of-completion revenue-recognition method while the completed contract method is used in Statement 2. Both methods are acceptable for accounting purposes. Furthermore, equipment is depreciated using straight-line depreciation in Statement 1 while an accelerated-depreciation method is used in Statement 2.

Statement 2 reflects the depreciation of the office building. Statement 1 assumes that the owner of the firm chooses to keep the ownership of the office building separate from the construction company. (Note: Regardless of whether the office building is held by the individual or the corporation, the fact remains that there is an $80,000 liability for the building.)

The difference in the material cost on the two statements reflects using FIFO inventory costing in Statement 1 and LIFO inventory costing in Statement 2.

The noteworthy point is that all of the above accounting methods are acceptable. By admission, the numbers used are selected to prove a point. However, other alternative accounting methods also result in statement differences. While the balance sheet (statement of financial position) is not shown, the alternative accounting methods would also have a major effect on this financial statement including equity ratios and debt ratios.

The moral of the story is that not all financial statements are equal. The availability of alternative accounting methods emphasizes the need for full disclosures of accounting policies and notes in the financial statements. Only then can the reader analyze and evaluate the financial structure of the construction firm.

# 7.4  Financial Ratio Analysis

Numerous financial ratios can be developed by dividing one of the many different items in a firm's income or balance statement by another item. The firm can then compare the values of its ratios to the average values of the ratios for its industry and perhaps decide to make an adjustment in its financial structure. Ratio analysis is also used by external parties in evaluating the financial structure and risk of the firm. Sureties and bankers will typically concentrate on what they view as key financial ratios when determining the feasibility of bonding the firm or lending money to the firm.

The financial ratio analysis user should be aware of its limitations. For one, each industry and firm is somewhat unique as to its product, its objectives, and its financial structure. It follows that each firm is somewhat unique as to the value of its ratios. Basing a firm's policies on a specific growth objective may result in its ratios varying from those accepted as average for the industry, although, the firm may have a very sound financial structure.

The optimal financial structure for a contracting firm can change over a period of years. What were ideal financial ratios ten or five years ago, may not be ideal today. Changes in interest rates, accounting practices, and tax laws are a few of the external factors that can justify changes in the financial structure and financial ratios of the firm.

Perhaps the greatest difficulty with ratio analysis is the fact that a firm can greatly manipulate its financial ratios through transactions that are for this purpose alone. For example, should a firm need to improve its short term position of current assets versus current liabilities, it can issue long-term debt and thus either raise cash (a current asset) or use the cash to reduce its current liabilities. Similarly, selling fixed assets such as plant or equipment will provide cash that in turn will improve the value of the current ratio to be analyzed by a potential creditor.

A difficulty somewhat unique to the construction industry's use of financial ratios has to do with the alternative accounting methods used by the construction firms. Using alternative accounting methods such as the completed-contract method versus the percentage-of-completion method results in different values for certain financial ratios.

Regardless of the problems associated with financial ratio analysis, they continue to be used by the contracting firm and external parties. Specific ratio values are not as important as what the ratios are supposed to indicate. Knowing that a given ratio is indicative of a firm's ability to meet payment on its current liabilities is more important than trying to memorize specific acceptable values of the ratio. In addition, observing the trend of a given ratio's value is useful in that it may flag a potential future financial problem. Observing the ratio as a function of time will also aid in detecting any attempt by the firm to manipulate its ratios.

Whereas certain financial ratios are useful for evaluating the short-term soundness of the contracting firm's financial structure, others are more relevant to evaluating long-term soundness. Thus, potential creditors considering lending short-term loans may be interested in one set of ratios, potential creditors considering lending long-term loans and potential investors of equity capital may be more concerned with a different set of financial ratios. Let us first consider some of the more commonly used financial ratios that relate to short-term financial analysis.

## 7.5  Short-Term Analysis

While not truly a financial ratio, the measure of a contracting firm's working capital to its dollar value of work being performed is indicative of its ability to meet commitments as to material purchases and labor payroll. Working capital refers to current assets that are readily transferred to cash. For the contracting firm, this consists primarily of cash and current receivables. As a general rule, the contracting firm should maintain a 10 to 15 percent ratio of working capital to its dollar workload. A value less than this enhances cash-flow problems.

Creditors focus on financial ratios that center on the relation of current assets to current liabilities and the timing with which receivables and inventory turn into cash. Short-term creditors are concerned with the immediate and short-term ability of the contracting firm to generate cash for the payment of loan principle and interest. The financial ratios indicative of the contracting firm's short-term liquidity or ability to pay are derived from the firm's balance sheet.

The current ratio is by far the most recognized short-term financial ratio. It is a test of the firm's solvency and indicates the firm's ability to meet current obligations. The current ratio of a firm is derived as follows:

$$\text{Current ratio} = \frac{\text{Current assets}}{\text{Current liabilities}}$$

Typically, lenders of money prefer to see a current ratio of 2 : 1 or even 2.5 : 1. However, it is difficult for even the most financially sound contracting firm to maintain such a high ratio. The contracting firm's high dependence on short-term credit results in a realistic current ratio of about 1.5 : 1. Any deviation lower than a 1.5 : 1 ratio should be viewed as an indicator of future cash flow problems.

Theoretically, the quick ratio or acid test is an even more severe test of a firm's solvency. Whereas the current ratio includes inventories in determining assets, they are excluded in determining quick assets. The quick ratio or acid test is derived as follows:

$$\text{Quick Ratio} = \frac{\text{Quick assets}}{\text{Current liabilities}}$$

As is true of the current ratio, the quick ratio is derived from line items appearing on the firm's balance sheet.

The quick ratio might be considered as a more critical analysis of a firm's ability to generate cash over a short period of time. Unlike cash, current receivables, and temporary investments, it is sometimes difficult to transform inventory into cash without absorbing a loss on the transaction. As to the contracting firm, the difference between the current ratio and quick ratio loses some of its significance because the contracting firm is not heavily invested in inventories. This contrasts with other manufacturing industries such as the automobile industry which has a high inventory investment in raw materials. A quick ratio for a contracting firm that is less than 1.25 : 1 or even more critically 1 : 1 should be viewed with alarm.

The receivable turnover and inventory turnover financial ratios are indicative of a firm's ability to convert receivables and inventory into cash. The ratios are calculated as follows:

$$\text{Receivables Turnover} = \frac{\text{Credit sales}}{\text{Average receivables}}$$

$$\text{Inventory Turnover} = \frac{\text{Cost of project}}{\text{Average inventory}}$$

Neither ratio is as meaningful in the construction industry as it might be in other industries. The billing and payment process typical of the industry dictates the contracting firm's collection of receivables. However, should the firm deal in credit sales such as small repair jobs, then the receivable turnover ratio may be indicative of the soundness of the firm's accounting practices as to receivables and it collection procedures. For the most part, the contracting firm's receivable turnover depends mainly on the type of owner for which the firm performs work and the duration of projects the firm undertakes. Perhaps the most meaningful use of the receivable turnover ratio for the contracting firm is for the firm to observe changes in its own ratio's value with the goal of implementing procedures that improve the ratio's value.

Like the receivable turnover ratio, the inventory turnover ratio is distorted by the fact that the typical contracting firm has a relatively small investment in inventory. Since each project is somewhat unique, it is often unwise for the contracting firm to build up an inventory of raw materials. However, it should be noted that an in-process project is in fact inventory to the contracting firm. In this context, the inventory turnover ratio can be used to analyze the firm's ability to complete projects on schedule and turn them into cash. However, it should be noted that the value of the ratio will differ significantly depending on the type of accounting method used to recognize project income and project work in progress.

# 7.6   Long-Term Analysis

Although the long-term creditor or investor has to be concerned about the firm's ability to meet its current obligations., it is also concerned about the firm's long range profitability. Expected income growth, ability to pay future dividends, and the ability to raise long-term funding are of the utmost importance to the long-term financial success of the firm. In great part, these long-term success factors are reflected by the firm's income statement.

Of the many financial ratios used for evaluating the long-term financial soundness of the firm, those most used and most relevant to the analysis of the construction firm are as follows:

$$\frac{\text{Net profits}}{\text{Annual volume}}$$

$$\frac{\text{Net profits}}{\text{Net worth}}$$

$$\frac{\text{Net profits}}{\text{Net working capital}}$$

$$\frac{\text{Annual volume}}{\text{Net worth}}$$

The net profits to annual volume, net worth, and net working capital ratios relate to the earnings rate of the firm as a function of annual volume, net worth, and net working capital, respectively. Net worth equals the owner's invested capital plus any retained earnings. The value of the net profit to annual volume is usually relatively small for contracting firms versus other industries. A typical range of financial ratio values for net profit to annual volume is 2.0 to 4.0 percent for contracting firms. This "profit margin" is relatively low compared to other industries. However, one must remember that this profit margin is a rate of return on the customer's money not on the contracting firm's investment. In that sense, the ratio or percent cannot be compared to a rate of return associated with equity investment.

The values of net profit to working capital and net profit to net worth for the contracting firm tend to be higher than similar ratios in other industries. Ranges of these ratios for contracting firms are approximately 10 to 30 percent and 8 to 22 percent for the net profit to working capital and net profit to net worth, respectively. However, the relatively large values of the ratios are due to the relatively small amount of working capital and net worth of the contracting firm rather than a large dollar net profit.

The ratio of annual volume to net worth measures the number of times net worth is turned over in a specific time period. It is indicative of utilization of owner's equity and too low a value may indicate overcapitalization in relation to volume of work performed. Because of the widely varying dollar values associated with types of work performed by different contracting firms, the ratio of annual volume to net worth can vary substantially from one firm to the next. As such, as is true of the other financial ratios discussed, the most meaningful use often centers around the firm's comparing the value of the ratio as a function of time. Such a comparison can often flag needed corrective financial adjustments.

In addition to the short-term and long-term financial ratios discussed, numerous other ratios have been used with less frequency to analyze the financial soundness of the contracting firm. Included in these are the ratio of fixed assets to net worth, the ratio of net income to project direct costs (i.e., profit margin on cost), and various ratios that are aimed at analyzing the contents of the construction firm's overhead.

## 7.7   Industry-Wide Financial Ratios

Numerous services and associations publish industry wide financial ratio averages. Some of these services and associations limit themselves to publishing ratios for a specific industry while others publish and compare financial ratios of average firms in several industries.

Examples of services that publish financial ratios for several industries include Robert Morris Associates and Dun and Bradstreet. Figure 7.2 illustrates example published financial ratio data for a contractor.

Many of the services publishing financial ratios list a range of ratios for a given type of firm. The data that is published is secured from numerous firms resulting in variations in the values of a specific ratio from one firm to the next. The variation is characterized by the range of the financial ratios reported.

In addition to stating a range of financial ratios and an average in Figures 7.2, the ratio values shown in these figures are classified as to the size of the contracting firm. By characterizing the ratios by the size of the firm, this can enable the firm to compare itself with a firm of approximately equal size. Comparing the financial ratios of a construction firm to ratios of a firm five times bigger or smaller would result in a meaningless or even misleading analysis.

| < $1MM | $1MM & < $10MM | $10MM & < $50MM | $50MM & > | All sizes | Ratios |
|---|---|---|---|---|---|
| 1.7 | 1.4 | 1.2 | 1.2 | 1.4 | |
| 1.2 | 1.2 | 1.1 | 1.0 | 1.2 | Quick |
| 0.8 | 1.0 | 0.9 | 0.8 | 0.9 | |
| 2.0 | 1.6 | 1.5 | 1.4 | 1.6 | |
| 1.4 | 1.3 | 1.2 | 1.3 | 1.3 | Current |
| 0.9 | 1.2 | 1.1 | 1.1 | 1.1 | |
| 0.2 | 0.1 | 0.2 | 0.3 | 0.2 | Fixed Assets/ |
| 0.4 | 0.3 | 0.3 | 0.4 | 0.3 | Worth |
| 1.0 | 0.5 | 0.6 | 0.9 | 0.6 | |
| 0.7 | 1.1 | 1.4 | 1.5 | 1.1 | |
| 1.8 | 2.1 | 2.4 | 2.1 | 2.0 | Debt/Worth |
| 4.5 | 3.5 | 3.8 | 3.7 | 3.7 | |
| 11.9 | 9.1 | 8.1 | 15.2 | 9.4 | Revenues/ |
| 6.3 | 6.6 | 6.6 | 6.7 | 6.5 | Receivables |
| 4.2 | 4.8 | 5.2 | 5.8 | 4.8 | |
| 5.7 | 10.0 | 11.6 | 16.0 | 9.3 | Revenues/ |
| 14.2 | 17.7 | 23.9 | 24.1 | 17.9 | Working |
| 30.0 | 34.5 | 50.7 | 55.1 | 43.2 | Capital |
| 3.3 | 6.2 | 6.8 | 7.7 | 5.7 | Revenues/ |
| 7.6 | 10.0 | 10.8 | 11.0 | 9.9 | Worth |
| 17.0 | 15.6 | 16.8 | 17.4 | 16.2 | |
| 36.4 | 36.9 | 42.3 | 42.4 | 38.2 | % Profits Bef |
| 14.0 | 17.7 | 23.7 | 29.8 | 19.1 | Taxes/Worth |
| 12.1 | 3.4 | 10.1 | 17.8 | 3.4 | |
| 15.6 | 11.5 | 10.4 | 14.2 | 12.0 | % Prof B Tax/ |
| 2.8 | 5.3 | 7.3 | 10.7 | 5.5 | Total Assets |
| 8.4 | 0.8 | 2.9 | 5.1 | 0.6 | |

**Figure 7.2.** Example financial ratios — contractor.

## 7.8   Trend Analysis

The difficulty associated with comparing any one contracting firm's financial ratios with published averages for the typical contracting firm was discussed earlier. In particular, each firm has its own unique characteristics, including financial characteristics, that limit the usefulness or validity of the comparison.

A meaningful evaluation of the contracting firm can usually be made by charting various "key" financial ratios of the firm as a function of time. This is commonly referred to as trend analysis.

Trend analysis of financial ratios can lead to early detection of a deteriorating financial structure or a weak link in the overall financial structure of the construction firm. By detecting a potential shortcoming early, the contracting firm may be able to make decisions to correct a problem or at least minimize its impact.

A worksheet for performing trend analysis on key financial ratios for a contracting firm is illustrated in Figure 7.3. Examples of data for a contracting firm are also shown in the figure.

| (000's) | X1 | X2 | X3 | X4 | X5 |
|---|---|---|---|---|---|
| Sales | 2,460 | 2,560 | 2,830 | 2,835 | 2,850 |
| % Sales growth | — | 4 | 10.5 | 1 | 1 |
| Labor | 800 | 820 | 780 | 810 | 900 |
| Material | 925 | 900 | 1,010 | 1,010 | 1,040 |
| Subcont. | 500 | 550 | 720 | 765 | 680 |
| Gross margin | 235 | 290 | 320 | 250 | 230 |
| Gross margin % | 9.5 | 11.3 | 11.3 | 8.8 | 8.0 |
| Gen % adm | 133 | 172 | 202 | 218 | 220 |
| G & A % sales | 5.4 | 6.7 | 7.1 | 7.7 | 7.7 |
| Income before tax | 102 | 118 | 118 | 32 | 10 |
| Income before tax % | 4.1 | 4.6 | 4.2 | 1.1 | 0.4 |
| Admin salary | 24 | 27 | 32 | 38 | 39 |
| Assets | 800 | 812 | 830 | 850 | 830 |
| Sales to assets | 3.07 | 3.15 | 3.4 | 3.33 | 3.43 |
| Equity | 350 | 287 | 285 | 280 | 205 |
| Sales to equity | 7.03 | 8.92 | 9.93 | 10.12 | 13.90 |
| Current assets | 520 | 532 | 475.8 | 510 | 511.5 |
| Investments | 100 | 110 | 120 | 110 | 120 |
| Property % eq | 280 | 280 | 354.2 | 360 | 338.5 |
| Accum depr. | 100 | 110 | 120 | 130 | 140 |
| Current liabilities | 325 | 380 | 390 | 425 | 465 |
| Long-term liabilities | 125 | 145 | 155 | 145 | 160 |
| Current ratio | 1.6 | 1.4 | 1.22 | 1.2 | 1.1 |
| Debt to equity | 1.28 | 1.83 | 1.91 | 2.04 | 3.04 |
| Inc to equity % | 29.1 | 41.1 | 41.4 | 11.4 | 4.8 |
| Inc to assets % | 12.7 | 14.5 | 14.2 | 3.8 | 1.2 |
| Rev to cur asset | 4.73 | 4.81 | 5.94 | 5.56 | 5.57 |

**Figure 7.3.** Trend analysis.

Inspection of the financial data in Figure 7.3 indicates that the firm's profitability has leveled off and actually decreased during the last two years. Even though volume has somewhat increased during the last two years, the increase in direct costs and general and administrative expenses of the firm offset the increased volume.

Could the contracting firm characterized by the data in Figure 7.3 have foreseen the difficulties of the last two years by analyzing its financial data and financial ratios? The answer to this question is in part yes. Inspection of the first three years of data illustrates that the firm's general and administrative costs were increasing at a rate greater than the firm's gross profit.

Given this fact, the firm's total profitability becomes very dependent on the ability to increase volume. Any small decrease or leveling off of volume can lead to a significant decrease or leveling off of gross profits that can lead to a significant decrease in bottom-line profits given a large overhead expense (general and administrative).

A second indicator of potential problems after the first three years that can be detected is the firm's increased dependence on liabilities. In effect these liabilities create a fixed drain on the firm's cash and total profits.

Another danger signal that should have warned the contracting firm of potential problems and no doubt was the cause for some of the problems was the negative trend of the firm's current ratio. From a value of 1.6 in year 1, it had decreased to a value of approximately 1.2 by the end of the third year. The contracting firm has to be attentive to keeping its current ratio favorable (ideally a value of 1.5 or greater) in order to obtain the bonding necessary to maintain or increase the firm's dollar volume of work.

What, if anything, could the contracting firm have done after three years to prevent the financial difficulties of the last two years? The firm probably should have attempted to reduce its general and administrative expense. Total profits depend on three factors: volume, direct expense, and overhead costs (general and administrative expense).

The reason the contracting firm was unable to increase its volume at a greater rate during the last two years is not clear. Undoubtedly part of the reason was its deteriorating current ratio and inability to maintain or increase bonding. However, the reduction in volume was probably also related to less demand for construction or more competition from competing contractors. The result is that the contracting firm is not in total control of its annual volume.

The fierce competition within the construction industry in great part dictates the firm's gross margin (shown as approximately 10 percent for the firm illustrated in Figure 7.3). In other words, the contracting firm's direct costs of performing construction work are not totally controllable by the firm. This is not to say that the firm's management cannot use cost-reducing construction methods. However, there are practical constraints including competition that limit the amount of cost savings that can be obtained.

The result is that the most controllable element effecting total firm profits in Figure 7.3 is the firm's general and administrative costs. Attention should focus on this cost because it has continued to grow at a rate faster than volume. More often than not, a contracting firm is non-attentive to excessive operating expenses during profitable years. The "fat" of the business tends to be overlooked because of the firm's significant profits. However, as soon as volume slows up or decreases, the excess operating expense has a multiplied negative effect on the firm's profitability.

As a minimum, the contracting firm should evaluate its overhead expenses annually. This analysis can establish required volume for profitable operations or lead to a necessary reduction in overhead to enable profitable operations.

Having calculated the break-even volume, the contracting firm could direct its marketing efforts to securing the required volume. If this is judged unfeasible, a reduction in operating expense is probably necessary in order to prevent an unprofitable year.

A reduction in overhead expenses may not have been the only management practice that could have improved the results of the last two years shown in Figure 7.3. As noted earlier, the deteriorating current ratio of the contracting firm probably leads to some of the problems that surfaced in years 4 and 5. To correct this negative characteristic, several options were open to the firm. First, the owners could have chosen to take less cash out of the firm as salary or as distribution of profits in order to enhance the current ratio. Also, some fixed assets could have been sold, or the firm could have curtailed some of its expenditures for fixed assets. Either of these two actions would have improved the firms current ratio. The improved current ratio may in turn have led to more volume and more profits.

In summary, many of the favorable or unfavorable financial results of the contracting firm do not happen over night. Analysis of the firm's financial statements and financial ratios for one year to the next can lead to early detection of a future event. This analysis, commonly referred to as trend analysis, can serve to assist the contracting firm in managing itself rather than being totally depending on factors it cannot control.

# 7.9   Concerns of Sureties and Lending Institutions

When evaluating their willingness to sell a construction firm a bond or give it a loan, both the surety company and the lending institution share common concerns. These include the following:

1. Integrity and personal character of the owners of the construction firm
2. Technical know how of the individuals
3. Experience and performance track record of the construction firm
4. Financial strength of the construction firm

The factors are not necessarily listed in order of importance. For the contracting firm seeking its first bond or loan, the integrity and personal character of the firm's owners will be especially important. For the ongoing firm, the first three factors become of secondary concern versus the analysis and monitoring of the construction firm's financial statements.

The importance of any of the factors in evaluating the contracting firm depends on whether the analysis is being made by a surety or lending institution. Surety companies tend to be less subjective in their analysis than lending institutions. The surety company often avoids making personal opinions. Instead, they stick to established formulas and factors when evaluating the contracting firm. This is not to say that factors other than financial ratio and statements are disregarded. Experience and performance along with other factors are weighted in the evaluation. However, the factors and weighting system employed by sureties tend to be more uniform in application from one contracting firm to the next compared with lending institutions.

Surety companies and lending institutions pay close attention to a few select financial ratios of the contracting firm. For example, surety companies generally prefer to limit performance bonds to approximately ten to eighteen times a contracting firm's net working capital and approximately ten times the firm's net worth. Modifications to these ranges are made for firms with good or bad experience, track record, and other financial statement characteristics.

Regardless of who is appraising the financial statements, the contracting firm's ability to carry out business with an external party is enhanced if the financial statements and resulting financial ratios are favorable. The high dependence of the contracting firm on external parties places even more importance on keeping the financial statements and financial ratios in good light.

## 7.10  Summary

The contracting firm's financial statements are often used by external entities as a basis of evaluating the firm. The surety company and lending institution review and evaluate various financial ratios of the firm.

There are many financial ratios that aid in the evaluation of the firm. The surety company and lending institutions focus on ratios that can be calculated from the balance sheet of the contracting firm. Included in these ratios are the current ratio, debt to equity ratio, and working capital.

While each contracting firm is somewhat unique in regard to their financial ratios, the firm can evaluate their own financial ratios as a function of time as a means of monitoring their operations. This process of evaluating financial ratios as a function of time is referred to as trend analysis.

# Chapter 8

# Business Law

## 8.1  Introduction

The subject of law is of such vastness that no one could possibly know all there is of it. A dictionary containing only definitions of legal terms is a substantial volume. Undoubtedly one cannot expect the contractor to be knowledgeable regarding all of the subdivisions of law such that he can consider himself a lawyer capable of seeing to all of his own legal interests. The services of a lawyer are sometimes necessary to the financial success of the contractor. However, the contractor may often avoid the need for services of a lawyer or the costs of a lawsuit by means of a general knowledge of circumstances that often lead to legal liabilities or suits. Such knowledge can reduce, if not eliminate, his very involvements in legal disputes and lawsuits.

The contractor receives no financial gain as a result of becoming involved in a lawsuit or dispute. At best, the suit or dispute is resolved in favor of the contractor, and he is placed in a position equal to that that would have existed had the dispute not arose. Obviously, the contractor would be just as well off had the suit or dispute not arisen. Not all lawsuits or disputes in which the contractor becomes involved are resolved in his favor. In addition to having to perform unexpected tasks or incur unexpected costs when a suit or dispute is resolved against him, the contractor will also be liable for court costs and legal fees associated with the judgment.

The various types of laws in existence in the United States can be classified as either common law or statute law. Common law consists of doctrines which have their origin in court decisions. These laws are the result of the judicial process. Statute law consists of rules of conduct enacted by authorized individuals. These laws represent the will of the lawmaking power and rendered authentic by enforcement of legal authority.

The subject of law is often divided in public law and private law. Private law is alternatively referred to as civil law. The public law pertains to

the public as a whole. Private law concerns itself with the relations between
individuals within society. Public law is divided into constitutional, administrative, and criminal law. Private law is divided into tort and contract
law. Contract law is undoubtedly the single most important field of law in
regard to construction contracting. However, contractor knowledge of his
liabilities in regard to all fields of law can reduce his risk of being involved
in a lawsuit.

## 8.2  Constitutional and Administrative Law

Constitutional and administrative law are part of public law. Constitutional
law deals with the power of Congress, the legislature, and the federal and
state governments. Two constitutions are in force in each state-one is the
state and the other is the national constitution. Constitutional law concerns
itself with such matters as the power to tax, bankruptcy laws, and the
establishment and regulation of currency.

Administrative law concerns governmental bodies which are created by
the Congress or state legislatures for the purpose of enacting what amounts
to further legislation. One of the distinguishing characteristics in American
law in the last half century has been the growth in number and importance of administrative law and administrative bodies and commissions.
An example is the National Labor Relations Board. Congress enacted a
law creating the board and granting it certain powers. The board in turn
makes laws.

## 8.3  Criminal Law

Criminal law is another important phase of public law. Here certain actions
by individuals and groups are prohibited and penalties are enforced in the
name of the public as a whole. Proceedings may be instituted by a public
official or body such as a grand jury or a police officer or some individual
not personally affected, in the name of the public as a whole. Suppose
that Jones murders Smith. Jones will be proceeded against by the state.
If the crime takes place in Illinois, an indictment (written complaint) will
be issued. The case is not entitled "The orphaned children of Smith versus
Jones". The indictment after reciting the facts will end with the assertion
that the killing was "against the peace of Smith" and it does not allege
that damage was suffered by such children due to the loss of the father.
Crimes may be punished by imprisonment, fine, removal from office, or the
deprivation of the right to hold public office.

Crimes are subdivided into two large classes, felonies and misdemeanors. Felonies are serious crimes, punishable by imprisonment. Among these
are murder, rape, robbery, and burglary. Misdemeanors are crimes of a less

serious nature such as traffic violations, trespass to real estate, vagrancy, and other offenses of a similar nature. Punishment is usually by fine only or short sentences in such corrective institutions as local jails or work houses.

Whereas a contractor may often become involved in a contract law dispute due to his ignorance of the law, his involvement in the violation of the criminal law is somewhat intentional. Examples of such criminal acts include the defrauding or misleading of the public, embezzlement, and the receiving of stolen goods.

Contractors are guilty of defrauding or misleading the public when two or more contractors commit collusion in their fixing of their bids to be submitted to an owner for the building of the owner's project. Such a practice is commonly referred to as "rigging" the bids.

A contractor may be given or he may purchase at a discount, material or equipment that has been stolen. Regardless of whether the material or equipment has been stolen by the individual from whom they are received, or from an individual who was not the individual who stole them, if the contractor is aware that it is stolen material or equipment he is guilty of receiving stolen goods. It is immaterial that the contractor does not know the owner or the thief.

A contractor who obtains material from a supplier by means of a worthless check may be guilty of obtaining goods by false pretenses. The question that determines guilt is whether the contractor intended to defraud the supplier that the check had value when he knew in fact that it did not. If such is the case, the contractor is guilty of the crime.

Forgery differs from the crime of obtaining goods by false pretenses in that it consists of the material alteration of an instrument which creates or changes the liability of another party. The most common form of forgery is the fraudulent making of a check by means of signing a false signature. The construction project building process usually initiates the production of many checks. The contractor may receive checks from project owners, architects, engineers, or even material suppliers. When such a check is lost or stolen, the contractor may be victim of forgery. To eliminate this possibility, the contractor upon receiving a check, should immediately place his bank endorsement stamp on the back side of the check.

## 8.4  Tort Law

Tort law along with contract law make up private law. Tort means "wrong", in this case a "private wrong". As a result, tort law is sometimes referred to as civil law. It is possible, indeed, it is true more often than not, that a wrong inflicted upon an individual not only wrongs the individual but society as a whole. Hence, a crime is usually a tort, and a tort may be a crime which is the subject of public law in the form of criminal law.

Let us go back to Smith's murder by Jones. We have already seen that Jones's action was an offense against the peace and dignity of the People of the State of Illinois. Jones's action was also a wrong to Smith and Smith's family. Hence, although Jones may be tried, convicted, and executed under criminal law, a branch of public law, he may not have heard the last of the consequences of his wrongful action. Smith's widow and children may sue Jones or Jones's estate under private law in what is called a tort action for damages suffered by the loss of support furnished by the deceased husband and father. Jones's action was not only murder under the criminal law; it was also a trespass to the person of Smith. Trespass may take the form of an unlawful invasion of the person ranging from murder to simple assault, trespass to real estate resulting from unlawful entry thereupon, or even injury to one's reputation by slander or libel.

The construction contractor is likely to become involved in a tort due to either an intentional act or due to his failure to recognize his legal responsibilities. A tort can arise when the contractor violates an obligation or duty created by law. When the contractor is guilty of breach of contract that causes a wrong or injury to other parties of the contract, he is guilty of a tort. Thus the contractor's act can be both a breach of contract and a tort.

A difficult question arises when it comes to an accident occurring to a child when the child trespasses onto the contractor's project. In general, the contractor is under no obligation (no absolute liability) to make the construction site safe for infant trespassers. However, he is liable for the injuries to these trespassers if his building site is ruled to be an attractive nuisance to the infant trespasser.

Most torts are the direct cause of an individual's negligence. In torts of this type the question of guilt is centered around whether the defendant acted with less care than a reasonable man given the circumstances. The definition or description of a reasonable man is for the courts to determine. Thus, when a contractor trespasses the land of a land owner when transporting material to his building site, the courts have to decide if he acted as a reasonable man would have given the circumstances.

## 8.5   Contract Law

By far the most frequent involvement of the contractor in a legal conflict occurs in the branch of law referred to as the law of contracts. The contractor enters into contracts with building project owners, material suppliers, equipment manufacturers and lessors, subcontractors, and labor. The complexity of contracts, especially contracts with the building project owners, enhances the possibility of contract disagreement, a breach of contract, and a lawsuit. The complexity of the construction contract with a project owner is due in part because of the unique nature of every project, the many types of work involved in a single contract, and the large dollar value of the single contract.

As is true of other conflicts of law to which the contractor is a participant, the contractor is almost better off if the conflict had never arisen. The contractor's awareness of contract law as it relates to his everyday making and carrying out of his contracts can often result in the savings of money associated with project expenses, breach of contract, and lawsuits.

A contract can be defined as an agreement between two or more competent parties for a legal subject matter, whereby one party exchanges an act for the promise or the act of another party. A contract between two parties creates legally enforceable obligations on each of the parties. Upon one party's failure to perform his obligation, the other party can sue and receive damages from the party who is guilty of breach of contract.

A contract that is binding and enforceable on both parties of the contract is referred to as a valid contract. A voidable contract is binding and enforceable on both parties but because of certain circumstances such as one party being a minor, the contract may be rejected by one of the parties. A void contract is not binding or enforceable on either party of the contract.

Although it is best to make contracts in writing since this helps to eliminate misunderstandings of the participating parties, most contracts can in fact be made orally. Certain types of contracts as dictated by the statute of frauds must be in writing. The contracts that must be in writing are contracts to sell interest in property, contracts that can not be performed within one year after the contract is made, a contract for the selling of goods with a value in excess of $500, and a contract to pay the debt of another. The statute of frauds affects the form of the contract that the contractor makes with a building project owner in that the duration of the work involved often exceeds a year duration. Thus, the contract must be in writing. It should be noted that even if the contract is for a duration less than one year, it is rare to find a contractor–owner construction contract that is not in writing.

Most contracts to which the contractor is a participant are expressed contracts. These are contracts whereby each of the parties of the contracts have made oral or written declarations of their intentions. A contract does not have to be an expressed contract to be enforceable. An implied contract is enforceable if one party offers services to the other party indicating that he expects to be paid for the services, and the other party accepts the benefits of the services. Implied contracts are often the rule for several professional services such as those offered by an engineer, physician, or lawyer. A contractor may become involved in an implied contract when he is forced to perform work for which the actual scope of the work is not completely determined and the cost of his services can only be determined upon completion of the work. Such a contract is referred to as a cost plus contract.

In addition to express and implied contracts, the courts may also enforce the obligations of a party when in fairness and good conscience a party has received goods or services for which he should pay or return. This is referred to as a quasi contract. An example of a quasi contract is the performing of house improvements by a renter. If the owner of the house was aware of the fact (perhaps by seeing them) that the improvements were being made and he did not prohibit the renter from making them, the courts will likely require the owner to pay the renter for the improvements.

A contract is classified as a bilateral or unilateral contract. The more common contract, the bilateral contract, is one where one party, the offerer, makes a promise to do something in exchange for a promise from the offeree to do something. Thus, when a contractor promises a potential building project owner to build the owner's project, for a promise from the owner to pay the contractor a defined amount, they have entered into a bilateral contract.

A unilateral contract is one where one party, the offerer, makes a promise to do something for an act from the other party, the offeree. There is no mutual agreement in an unilateral contract in that the party who may perform the act is not legally required to perform the act. Thus, when a potential building project owner promises a contractor $30,000 if the contractor will build his project, the contractor can claim the $30,000 upon building the project. However, the contractor does not have to build the project in that he has not legally entered into an agreement to build it. Unilateral contracts tend to be "weak" contracts in regard to the offerer's legal remedies.

## Contract Agreement

In order to be a valid contract there must be both an offer and an acceptance. An agreement is initiated when one person, the offerer, makes an offer to another person who is referred to as the offeree. An offer may be expressed in writing, by words, or it may even be implied.

In order to qualify as a legal offer, the offerer must intend to create a legal obligation. If there is no contractual intention, there is no legal offer. A project owner's advertisement for contractor bids is merely an invitation to negotiate. Because of the lack of intention the advertisement does not constitute an offer. On the other hand, a contractor's bid for the project is an offer. However, the bid by itself is not a contract until there is an acceptance by the owner.

One is not legally obligated to carry out an agreement to make a contract at a future date. Thus, there is no legal contract when a contractor promises to enter into a contract at some future date to build an individual a construction project.

An agreement to enter into a contract at some future date differs from an option contract. An option contract is a promise to keep open an offer for a stated period of time or to a specific date. For example, a construction contractor may give an equipment finance company $100 for the option of purchasing a certain piece of equipment from the company within a 30-day period. The equipment finance company cannot revoke its offer to sell the equipment during the option period. The $100 the contractor paid for the option generally goes toward the payment of the asset upon purchase. If the contractor decides not to purchase the asset, a portion or all of the $100 is forfeited to the equipment company.

Whereas, an offer usually has to be definite and certain in substance in order to constitute a valid offer, there are exceptions. For example, a cost-plus-construction contract is valid even though the offer made by the contractor is not certain in regard to the dollar value of the cost of the work. Such contracts are ruled valid in that the type of work involved (uncertain work quantities) necessitates such a contract. The courts acknowledge the fact that to require a definite dollar amount (a lump sum contract) would result in the contractor bearing an excess amount of financial risk associated with work and cost uncertainties.

An offer to contract can be revoked by the offerer before it is accepted by the offeree. Of course, this privilege does not hold in regard to an option contract. If an offer is accepted by an offeree before he is notified that the offer has been revoked, the contract is valid. Thus, if a contractor mails his acceptance of an owner's offer to pay the contractor $50,000 for the contractor's building of a project, the contract is valid even if the owner mails his intent to revoke the offer before the contractor mails his acceptance. The owner's letter of revocation must be received by the contractor before the contractor's acceptance if the offer is to be legally revoked.

An offer is terminated upon a counteroffer by the offeree. Thus, an offer by a contractor to an owner to build the owner's project for $50,000 is terminated when the owner replies he will give the contractor $45,000 for the building of the project. The counteroffer becomes an offer made by the owner. It is now up to the contractor (now the offeree) to accept or reject the offer, or to make yet another counteroffer.

An offer may or may not include a date on which it expires. If a date is included, the offer terminates on the stated date. If there is no stated date, the courts will rule that the offer is for a "reasonable time". A reasonable time is the time a reasonable man acquainted with the type of proposed contract would assume it to be, given the circumstances. For example, a contractor's offer to build an owner's project would be judged to terminate at the letting of a project.

In addition to requiring an offer, a valid agreement must contain an acceptance by the offeree. The offer can only be accepted by the person to whom it is intended. The acceptance by the offeree can be in any format which expresses the intent to accept. Thus, a contractor's acknowledgment by saying okay over the telephone to an offer constitutes an acceptance. Although either the offer or acceptance can be made in such a haphazard manner, it is advantageous to have a more formal acknowledgment in order to avoid disputes.

An acceptance must be unconditional. If a change of the offer is made by the offeree, a new offer is in effect made. An offeree's failure to respond to an offerer's offer does not constitute an acceptance. Thus, an owner's failure to respond to a contractor's offer to build a project for a stated sum of money does not result in any owner legal obligations.

The validity of a contract may be affected by a mistake by one of the parties involved in the contract. A mistake in a contract may be a mistake of fact or a mistake of law. It may also be a unilateral or a bilateral mistake.

If there is a mutual mistake of fact by the parties the contract is normally ruled to be void. For example, if a contractor enters into agreement with an owner for the remodeling of the owner's building for a stated sum of money, and the building has burnt down and neither party is aware of the fact, the contract is void.

Ordinarily a unilateral (one party) mistake of fact does not affect any of the legal obligations imposed by a contract. However, there are exceptions. For example, a contractor may mistakenly add his bid items resulting in him submitting a very low bid price to an owner. If the mistake is so apparent to the owner that he should be able to recognize the understated price, the courts will usually rule that the owner cannot hold the contractor to the bid price. Even though the courts recognize that negligence of the contractor is not grounds for invalidating the contract, they also recognize that the owner should not take undue advantage of the contractor.

An agreement is not valid if one of the parties concealed information that would have influenced the other party not to enter the agreement. However, if the complaining party should have been aware of the concealed facts, it is not the obligation of the other party to bring them to his attention. A party to a contract is expected to know information available to a reasonable man. Thus, it is not the responsibility of an owner to inform a contractor of the soil conditions underlying a proposed building. However, if the soil is of extremely abnormal condition such that the contractor would not be expected to know of its state, the owner has the responsibility of informing him of the conditions. His concealment of the fact may be grounds for judging the contract void or voidable (the contractor may choose to accept the contract regardless of the concealment).

Fraud is yet another reason for a contract not being valid. Fraud occurs when a person misrepresents a fact, known or believed by him to be untrue, with the intention of causing another party to enter into a contract. Besides being grounds for making a contract invalid, fraud is a tort. The injured party can recover money damages resulting from the fraud of another.

The validity of a contract is removed if one party causes another party to enter a contract under duress. Duress relates to the depriving of an individual's free will as a result of a threat of violence. A threat of economic loss is not grounds for duress. Thus, a threat from an owner not to provide building material credit from the owner's supply subsidiary is not considered duress in influencing a contractor to build a project for the owner.

A contract which involves concealment, fraud, or duress is considered to be a voidable contract. The offended party may choose to accept the contract or he may choose to avoid it. If he accepts it, the contract is then valid. If he avoids the contract (referred to as rescinding the contract) he is required to restore the other party to his original position. That is, he cannot take financial advantage of the other party. Of course, if he has been the victim of fraud, he has a claim to damages.

As mentioned earlier, certain contracts have to be in writing. A construction contract agreement for which the work is to take more than one year is such a contract. If the contract violates the requirement that it be in writing, it may be judged void or voidable, depending upon the statute of the state in which it is made.

In many cases disputes arise as to what was offered and agreed to in a contract. Usually the courts will decide the dispute by means of studying the written agreement. If there is a written agreement, the courts will not allow contradictory testimonies of one of the parties. This type of testimony is referred to as parol evidence. Parol evidence is only permissible when there is evidence of fraud, or the contract is not complete and therefore needs clarification.

The courts will interpret an agreement as it is stated or implied. Clerical errors and omissions are ignored and the contract is interpreted as the courts judge its intent. If there are contradictions between what is written and what is typed in an agreement, the written part is ruled to prevail. A conflict between a numerical quantity that is written in words and written in figures is resolved in favor of the amount stated in words. When a contractor makes a contract in one state and performs the work stated in the contract in another state, the contract is often written with a clause indicating the state to which laws are to apply.

## Competent Parties

One of the requirements of a valid contract is that the parties to the contract are legally competent parties. The most common examples of parties lacking authority to enter into a valid contract are minors, intoxicated persons, and corporations acting beyond their legal powers (ultra vires act).

When a contract is signed between a legal party and a party lacking the legal capacity to enter into a contract, the agreement is normally a voidable contract. In the case of a corporation acting beyond their powers, rather than rule the contract invalid, the courts will neither enforce the contract nor hold either party liable for any breach of contract.

The party lacking legal competence to enter a contract has the option of avoiding the contract (referred to as disaffirmance) or he may perform his voidable contract (ratify it). The competent party does not have the right to disaffirm the contract. The person lacking competence (e.g., a minor) upon disaffirmance of the contract usually has to set things back to their original position. That is, he must return what he has received as a result of the contract. This act of returning consideration received is referred to as restitution.

Many of the states have statutes that require certain professionals to obtain a license in order to practice their profession. This license may be required in order to protect the public from individuals who are unqualified. On the other hand, several states impose a license requirement merely as a means of raising revenue by requiring a fee for the license. If the license is required for the purpose of protecting the public, a party practicing without a license is in fact entering a voidable contract when making a contract for such services. For example, if a contractor does not have such a required license when entering an agreement to build an owner's project, the contractor may find it difficult to win contract disputes against the owner. The owner has the option of avoiding or performing such an agreement. However, once the owner indicates that he is going to carry out his contract obligations, the voidable contract becomes a valid contract.

If the license is required for the purpose of raising revenue, the contractor's failure to obtain the license has no legal bearing on the validity of his contracts. Obviously, it is not always clear whether a license is required for public protection or merely for raising revenue. For this reason and for reasons of maintaining good external relations, it is rare to find a contractor who practices without a required license.

In regard to the parties of a valid contract, more than two parties can be involved. That is two or more parties may make a contract with one, two, or more than two parties. When two or more parties jointly promise to perform an obligation, or when two or more parties are jointly the recipient of benefits from an agreement, the contract is referred to as a joint contract. The construction contractor enters into such a contract when he combines his labor force and equipment with those of another contractor for the purpose of building a construction project. Such an arrangement is referred to as a joint venture. Such an arrangement is somewhat similar to a partnership. The main difference is that the joint venture is usually for a single project, whereas a partnership is formed for the carrying out of business for an unstated period of time.

Both of the parties to a joint contract must be legally competent parties. Any court action on a joint contract must be brought against all of the joint parties.

A third party beneficiary contract is a contract where two parties enter into an agreement whereby one of the parties is to perform an obligation for a third party. Such a contract is a valid contract. An example of such a contract is where a contractor signs an agreement with an owner where the owner promises to pay the contractor a sum of money for the promise of the contractor to build a project for a creditor of the owner. The third party, the creditor, has a right to sue the contractor for breach of contract upon the contractor's failure to carry out his obligations.

Not all contract beneficiary parties have the right to sue for breach of contract. Such is true in the case of an incidental beneficiary. A landowner does not have the right to sue a contractor upon the contractor's failure to carry out his obligations. This issue of beneficiary contracts is especially pertinent to public projects.

For example, a landowner does not have the right to sue a contractor upon the contractor's failure to properly pave a city street along which the landowner's property lies. Only the city can sue the contractor for his lack of carrying out his obligation.

The question as to whether the third parties harmed as a result of a party's negligent performance can sue the party is a difficult one to answer. In several states, only the parties directly involved in a contract can enter a breach of contract suit. Thus, if a contractor negligently fails to take measures to make a project safe for which he has contracted with a potential owner, a third party injured because of that failure cannot sue for breach of contract. However, it should be noted that a third party injured due to a contractor's negligence can often succeed in obtaining damages in a tort action centered around the contractor's absolute liability.

A competent party to a contract can transfer his rights from the contract to another party outside of the contract. Such a transfer is referred to as an assignment of contract rights. The party to the contract who makes the assignment is referred to as the assigned, and the recipient of the rights is referred to as the assignee.

A contractor can assign his right to receive payment from a building owner to the bank. This is often done for the purpose of providing security to the bank for a loan. The contractor is the assignor and the bank is the assignee.

Certain rights or obligations to a contract cannot be assigned regardless of whether the parties involved are legally competent or legally incompetent. A right to have an employee work for oneself cannot be assigned. If one enters into a contract for the unique services of an individual (e.g., the services of a well-known architect), the individual cannot assign his obligations to perform. An assignee of a contract can sue for breach of contract as if he were a party to the contract.

When a construction contractor who is a party to a contract to build a project for an owner is subjected to grave financial difficulties, the owner may agree to release the contractor from the contract. The owner will then seek another contractor to take the place of the original contractor. This type of contract change is referred to as a novation. The old contract is substituted with a new contract.

## Legal Subject Matter

In order for a contract agreement to be valid, the formation and the performance of the agreement must be legal. Thus, the agreement must be for legal subject matter. If it is not, the resulting contract is usually void. However, if the circumstances of the agreement are such that an innocent party to the contract is subjected to an unfavorable position as the result of the illegal act of the other party, the contract may be voidable. In this case, the innocent party may choose to avoid the contract or he may choose to have the contract performed.

There are several types of agreements that do not meet the requirements of being for legal subject matter. Included in these are agreements calling for an act that is a crime, an agreement that involves a conflict of interest (e.g., government officials are prohibited from entering contracts that relate to their public position), and agreements requiring public service (e.g., the payment of a public official for something he is required to do).

In regard to agreements which the contractor may enter, several practices may constitute illegal subject matter. In some cities and states there are laws that regulate the business activities of certain businesses. For example, in the previous section the licensing of contractors was discussed. The issued license often regulates the activities of the contractor. If the contractor violates the stated regulations he may be subject to a fine or criminal prosecution.

A contract in restraint of trade is another type of contract that is not valid due to illegal subject matter. A contractor cannot enter a contract with other contractors to create a monopoly on the construction market, or to "fix" bid prices on a given project. In addition to making the contract void, such practices are subject to civil and criminal penalties. Such contracts violate fair trade agreements. A similar violation occurs when the contractor performs below cost for the purpose of harming or eliminating his competition. Although several states have statutes against such a practice, the vagueness of the term below cost results in the statutes being relatively ineffective. In addition, the laws are hard to enforce in the case of the construction competitive bidding procedure in that the very procedure promotes such an "undercut competition" practice.

Certain contracts which the contractor enters into with a project owner may also be invalid due to illegal subject matter. For example, a contract to build a project which violates a city zoning law is an agreement that is not valid because of the illegality of the subject matter. Such contracts are ruled to be void.

A contractor is often involved in the borrowing of money for the payment of project and equipment expenses. State statutes regulate the maximum interest rate which a person or lending institutions can charge borrowers for the use of their money. If the person or institution charges an amount in excess of this specified amount they are guilty of usury. If a party is guilty of usury, the loan agreement is not valid due to the lack of legal subject matter. The remedies that are open to the borrower as a result of the lender's usury vary from state to state. Some states require the borrower to perform the loan agreement, adjusted to a legal interest rate. Others declare the entire loan agreement void and require that both lender and borrower be placed in their original position before the loan. Still other states penalize the lender who is guilty of usury by allowing the borrower to keep the borrowed money without requiring him to pay the money (principle or interest) back to the lender.

## Adequate Exchange

In addition to the agreement, legal parties, and legal subject matter requirements of a valid contract, there must also be an element of exchange. Each of the parties to the contract must exchange something with the other party. This exchange is often referred to as the providing of consideration. Consideration is what one promises or does for the promise or act of the other party. This promise or act may be in the form of money, physical property, a service, or even a promise not to do some act (referred to as forbearance).

If only one party offers consideration while the other party does not, there is no valid contract. For example, if a contractor merely offers to build an owner's project there is no contract in that the owner has not offered consideration.

The validity or the enforceability of a contract is not affected by the inadequate nature of one party's consideration. For example, if a contractor offers to build an owner a project for $20,000 from the owner, the fact that it will cost the contractor $25,000 to build the project does not make the contract valid.

A difficult question in regard to a contract's requirement for consideration arises when a contractor requests and obtains additional money over and above the amount stated in the original contract. This requesting of additional money is often related to work covered by the original work. For example, a contractor may contract with an owner to build a project for $20,000. Upon completing 75% of the work and having absorbed $20,000 of costs, the contractor might inform the owner that because of high costs he will require another $5,000 to complete the project. The owner may agree to the extra $5,000. However, upon further thought the owner may decide not to pay the additional $5,000. His defense is based on the fact that his $5,000 promise was in exchange for work covered in the original work and he therefore received no consideration for his $5,000 promise. Technically the owner is correct and the courts may rule in his favor. However, several courts have ruled in favor of the contractor (entitling him to the additional $5,000) on the grounds that the original contract was rescinded (done away with) when the owner agreed to the additional $5,000.

It should be noted that if the owner had not agreed to the additional $5,000 that no legal dispute would have arisen in that the contractor would have to absorb all the costs, whatever they may be, as part of the risk of the contract. It should also be noted that if the reasons for the contractor absorbing added costs were not in the control of the contractor (e.g., extremely unseasonal weather) the courts may also rule that the contractor is entitled to extra money regardless of whether the owner agreed to the claim. Such a ruling is centered around the fact that the conditions that prevailed during the construction were not those implied in the contract. However, there is a difficult question here as to what conditions the contractor should expect. In many cases the courts rule that the unforeseen conditions are part of the risk of contracting and costs associated with them are to be absorbed by the contractor. The fact that the contractor cannot obtain additional payment can also be explained by the fact that the courts do not enforce extra payment merely because a party did not obtain what he expected.

A different situation arises when the contractor agrees to accept $950 before it is due in payment of the $1,000 in full. In this case, the owner does give new consideration in that he makes early payment. Such an agreement does meet the requirements of a valid contract.

## Contract Termination

The majority of contracts are terminated by the performance of the terms of the contract. However, circumstances may arise that result in a contract being terminated by impossibility of performance, by agreement, or by acceptance of breach of contract.

The construction contract is normally terminated when the contractor finishes building the owner's project according to specifications and in return receives final payment from the owner. If a dispute arises as to whether each of the parties has performed his obligations, the party claiming his performance must show that he has in fact performed.

The construction contract may or may not state a time period for performance. If it does state the time period, then performance should be made within the time period. If the performance is not completed within the time period, the courts may rule it a breach of contract. However, if it can be shown that the other party did not incur harm due to the delay in performance, the courts will often ignore the performance time period clause of the contract.

The contract may stipulate that a certain amount of money is to be paid by the party guilty of not meeting a performance date. This is often referred to as liquidated damages. These liquidated damages should not be construed as penalty costs. That is, liquidated damages are enforced only to the extent of the costs the innocent party incurs as a result of the other party's lack of performance by the stated date. Penalty clauses are void, liquidated damage clauses are valid.

If a stated time period for the performance of a contract is not stipulated, the courts generally rule that performance must be made within a "reasonable time". A reasonable time is defined as the period of time in which the type of contract in consideration is normally performed.

The rule of substantial performance often applies to construction contracts. This rule states that is the contractor performs all of his obligations, except for a slight defect in his performance that he did not make willfully, the owner cannot hold back the entire payment from the contractor. The contractor is ruled to have made substantial performance and is entitled to the entire payment due on the contract minus the amount it will cost the owner to cover the defect. Theoretically the rules of contract would hold that the contractor is entitled to no payment in that he has not performed his contract obligations. However, the courts recognize the unfair nature of such a performance rule in regard to the building of a construction contract. Thus, if a contractor contracts to build a project for $50,000 and performs his obligations except that the quality of a certain phase of the work is unacceptable and will cost the owner $500 to repair, the contractor is in fact entitled to $49,500 on the grounds that he has completed substantial performance. If it can be shown that the contractor intentionally performed negligently in regard to the $500 of unacceptable construction he has no claim on the owner. It should be noted that the intent is often difficult to prove. Thus, the courts usually rule in favor of substantial completion.

The acceptance of the contractor's performance of a construction contract is often subject to the approval of an architect or engineer who is hired by the owner. Unless it can be shown that the architect or engineer is guilty of fraud, the acceptance or rejection of the performance of the contractor by the architect or engineer is final and binding on both the contractor and the owner. When the contractor deals directly with the project owner, the contract is often written that the contractor's performance is subject to the satisfaction of the owner. In such a case, the courts rule that the owner must pay the contractor according to the contract if the contractor's work would be acceptable to a reasonable man.

A contract is not discharged on ground of impossibility of performance merely because one party finds it an economic burden to perform. Impossibility of performance is grounds for termination of a contract only when it is indeed impossible to perform. For example, if a contractor contracts with an owner of a building to repair the building for a stated sum of money, and the building burns down before the contractor's performance, performance is deemed impossible. In such a case, the contract is ruled to be void.

Changes in laws which result in the contractor's performance being more expensive are not grounds for his nonperformance. For example, a new safety law may require the contractor to provide additional shoring for a building project. The fact that the law was nonexistent at the time the contractor entered into an agreement to undertake the work is not grounds for the contractor's nonperformance.

If a party to a contract is constrained in his performance of the contract then the innocent party is relieved of his obligations on the grounds of impossibility of performance. Thus a subcontractor is relieved of his obligations of a contract with a general contractor when the general contractor does not perform the work precedent to the subcontractor's performance.

Impossibility of performance on grounds of bad weather (referred to as acts of God) is no defense. That is, bad weather is part of the risk of contracting. Modern day construction contracts often provide for extra time for the contractor when he is subjected to such conditions. Even when such a stipulation is not in the contract, when abnormal conditions persist the courts sometimes rule that there is not breach of contract and extend to the contractor extra time to perform.

The two parties to the construction contract may terminate a contract by agreement. For example a project owner may find it difficult to raise the money needed to initiate construction, and the contractor realizing that he may get future work from the project owner if he cooperates, may decide to agree to terminate the contract agreement. The government often reserves the right to terminate a construction contract at any time during its performance. An agreement of this type with a contractor usually provides for liberal termination benefits for the contractor should the owner decide to terminate the project.

Because of changes of work units, the contractor and the owner may agree to a different performance than that stated in the original agreement. Such an agreement is referred to as accord. When the new agreement is performed there is accord and satisfaction and the original contract is disregarded. To avoid the requirements for an entire new contract every time work changes occur, construction contracts are usually written to provide for change orders. The contract states that the contractor will be paid an amount in addition to the stated contract sum for work changes that occur. Usually this is in the amount of the contractor's cost plus a stated profit.

When one party to the contract fails to perform his obligations of the contract, he is guilty of breach of contract. Upon the breach of contract of one party, the other party is relieved of his obligations. However, if the contract term broken is not a sufficiently important part of the contract, the innocent party is not relieved of his contract obligations.

If the innocent party does not take action when the other party commits a breach of contract he is said to have waived the breach. In such a case, the innocent party cannot at a later date take action because of the other party's breach of contract. For example, if a contractor commits a breach of contract as a result of performing work that does not meet specifications, and the owner accepts the work, the owner cannot seek remedies at a later date for the contractor's breach.

An injured party is entitled to sue for damages upon the other party's breach of contract. If the injured party does not actually sustain a loss because of the other party's breach he is still entitled to nominal damages. Nominal damages are a small sum of money, such as a dollar or two.

If the innocent party to the contract sustains damages because of the breach of the other party, the innocent party can recover the damages. These damages are referred to as compensatory damages. If a contractor is guilty of breach because of inferior work, the owner is entitled to compensatory damages from the contractor equal in amount to the costs the owner incurs to hire someone to repair the work such that it meets specifications.

The courts will usually not require the breaching party to pay punitive damages. These would be damages in excess of those actually incurred by the innocent party and imposed for the purpose of punishment.

An innocent party is required to mitigate damages when he becomes aware of the other party's breach. This means that he cannot let damages increase if he can prevent them. If he does, he is not entitled to sue for such damages. Thus, a contractor cannot be entitled to those damages that he incurs on a building project after he is informed by the owner that the owner will not be able to meet further financial obligations.

In addition to being liable for damages to the owner when he fails to perform his obligations, the contractor may also be liable for a tort to a third party when he commits a breach of contract. Thus, when a third party is injured due to the negligence of the contractor when breaching a contract with an owner, the third party may bring a tort action suit against the contractor. Tort was discussed in an earlier section of this chapter.

Related to the performing of a contract and the breach of contract is a legal device referred to as a lien. A lien is a legal claim against the property of another for the satisfaction of a debt. There are many types of liens that are used in regard to the building of a construction project. These liens

that are drawn for use in the construction industry are often referred to as
mechanics liens. All lien laws are statutory laws.

In regard to the performance of the project owner and the construction
contractor, the contractor may obtain a personal property lien or a real
property lien. Thus, upon nonperformance of the owner's payment of the
debt owed to the contractor, the contractor may state claim to the personal
property or real property stipulated in the lien. In the case of a construction
project lien, the real property involved is usually the project itself. As is
true in any legal dispute, the project owner has a right to defend himself
in a lien dispute.

The first step in the establishment of the mechanic's lien is the filing
of a notice of mechanic's lien. It is filed with the clerk of the county in
which the property is located. The notice gives the name of the lienor (i.e.,
the contractor), the name of the project owner, a description of the labor
performed, and the materials furnished to the project. In addition it states
the contract price, the amount unpaid, the date of commencing the work,
and a description of the property. The time in which a notice of lien must
be filed by the contractor varies from state to state. A typical time is three
months. Once the lien is filed it must be acted on by the contractor within
a stipulated period of time. While the time varies from state to state, a
typical maximum time in which the contractor must act on his filed lien is
three years.

The lien rights of contractors and subcontractors are dependent on
whether the state in which the work is performed abides by the "New York
System" or the "Pennsylvania System". Under the New York System a
subcontractor is limited in the amount he can collect by the amount due
the general contractor from the owner. Let us assume that a project owner
enters into a contract with a general contractor for $20,000, and the general
contractor hires a subcontractor to do some of the work for $4,000. Under
the New York System, if the subcontractor gives the project owner notice
as to the amount of money that will become due on his contract with the
general contractor, then the project owner is entitled to withhold that sum
from any payments to the general contractor. If no notice has been given
from the subcontractor to the project owner and the project owner pays the
entire contract sum of $20,000 to the general contractor, the subcontractor
cannot collect his $4,000 from the project owner but must look to the
general contractor.

Under the Pennsylvania System the subcontractor has a right to file a
mechanic's lien for his labor and material even though the entire contract
price has been paid by the project owner to the general contractor. Thus
the owner may be forced to pay for the contract work twice should the
general contractor fail to make payment to the subcontractor. To avoid
this possible double payment, the owner will usually require a waiver of lien
when making payment on the work performed. Once the contractor signs

the waiver of lien, he no longer has any claim to ownership of improvements.

A contractor's lien on the project owner is not the only type of lien that is used in the construction industry. Others include a materials supplier's lien on the contractor and a subcontractor's lien on the contractor.

## 8.6 Commercial Paper

Commercial paper plays a role in the everyday business of the construction industry. Commercial paper consists of written promises or orders to pay money that is transferred by negotiation. Written promises to pay are usually in the form of promissory notes. Checks and drafts are examples of written orders to pay a stated amount of money.

Seldom does today's businessman pay his bills and receive revenue in the form of cash. He is much more likely to handle business transactions by means of commercial paper. Commercial paper serves as a substitute for cash. Commercial paper transactions often prove to be more convenient and safe in regard to the possibility of loss than are cash transactions. This is especially the case in regard to the construction contractor. Daily he engages in the writing and receiving of checks, and the writing of promissory notes and drafts.

Commercial paper transactions are similar in structure to contract agreements. However, the assignee of commercial paper is subject to less risk and given more rights than those given the assignee of contract rights. That is, commercial paper can be more easily assigned and rights transferred from one individual to the next. Commercial paper law is dictated by the Uniform Commercial Code.

A person who holds commercial paper that is payable to the individual holding the paper (his name does not have to be on the paper) is referred to as the bearer. This type of commercial paper is referred to as bearer paper. On the other hand, some commercial paper is payable to an individual only if the individual's name appears on the paper. Such paper is only payable to the order of the individual designated.

An owner of commercial paper can assign his paper to another individual by means of signing his name to the back of the paper. The person who transfers the paper by means of his signature is referred to as the endorser. The person to whom the endorser assigns the commercial paper is referred to as the endorsee.

An endorser can transfer commercial paper by means of one of several types of endorsements. A blank endorsement contains only the name of the endorser. There is no designated endorsee. As such, any individual who might find the paper holds bearer paper and can in effect cash the paper. Thus, while the blank endorsement remains the most commonly used endorsement, it is also the most dangerous in regard to risk of loss.

A less risky form of endorsement takes place when the endorser designates an endorsee. He would do this by writing "Pay to Jim Smith" on the back of the commercial paper. This type of endorsement is referred to as a special endorsement.

An even less risky endorsement is one where the endorser signs his name and adds the words "without recourse". Such an endorsement is referred to as a qualified endorsement. The effect of a qualified endorsement is that it relieves some of the liabilities of the endorsement in regard to the default of the drawer of the paper.

A restrictive endorsement is one that states the purpose of the endorsement. For example, a member of the ABC Construction Company would endorse commercial paper by means of a stamp that might read:

> For Deposit Only
> ABC Construction Company

The effect of such an endorsement is that the paper can only be transferred to the account of ABC Construction Company. However, a restrictive endorsement does not prohibit further transfer of the instrument.

A person who adds his signature to commercial paper and adds a statement that he will pay the value of the paper if the maker does not pay is referred to as a guarantor. This may often be done to add strength to the maker such that he can negotiate the paper. However, the guarantor places himself in a state of risk in that upon the maker's inability to pay (i.e., he is ruled to have insufficient property to cover the debt); the guarantor becomes liable.

In order to be negotiable (i.e., be transferable from one individual to the next), commercial paper must conform to certain format requirements. A fundamental requirement is that the paper be in writing and signed by the maker. The paper may be handwritten, typed, or printed. The maker usually signs his name to the lower right-hand corner of the commercial paper. However, this is not a requirement in that the paper is negotiable as long as his signature appears on the paper. In absence of a statute that forbids it, the initials or symbol which identifies the maker is a substitute for an individual's full signature.

Commercial paper must be unconditional in order to be negotiable. Thus, when an owner writes a check to a contractor and stipulates on it that it is payable upon the contractor's completion of construction work, the paper is not negotiable. The same is true in regard to the contractor's writing a promissory note to a bank for a loan in which he promises to make payment of the note upon winning a certain construction contract.

The amount of payment indicated on commercial paper must be certain in order to be negotiable. If the stated amount in writing on the paper differs from the amount printed on the paper, the written amount is ruled to be the amount due. Commercial paper must also be payable in money in order to be negotiable.

The paper should indicate a date or a time interval in which payment must be made. If it states that it is payable when a certain even occurs, it is non-negotiable in that the event may never occur. Certain types of commercial paper are payable on demand.

In order to be negotiable, commercial paper must be payable either to a stated individual or to any holder of the paper. When it is payable to a given individual it is referred to as order paper. If it is payable to any holder it is referred to as bearer paper. The paper may stipulate that it is payable to both order and in bearer. For example, it may state "Pay to Jim Smith or Bearer".

In addition to the general formal requirements for negotiability, there are additional legal terms and obligations that relate to specific types of commercial paper. These will now be discussed.

## Checks

The most used type of commercial paper in the construction industry (or any other industry) is the check. A check is an order by a depositor of a bank upon the bank to pay a stated sum of money to another individual. The individual who writes the check is referred to as the drawer. The individual to whom the check is addressed is referred to as the payee. The depositor's bank upon which the check is drawn is referred to as the drawee.

A check is in effect a kind of draft. However, it differs from a draft (as one commonly refers to a draft) in that the drawee of a check is always a bank, whereas in a draft the person to whom the draft is given is the drawee. In addition, due to state bad check laws, a violator of check laws may be subject to criminal action as well as being civil liable as in the case of a draft.

Although several banks' checks vary in style, most allow space for the drawer to note the purpose for which the check is written. It is beneficial for the drawer to place the purpose of the check in that the courts will recognize the written notation should the purpose be later questioned. In addition, the drawer can use the notation space to indicate that the check is for "Payment in Full". Upon the drawee's acceptance of the check, the corresponding debt is discharged regardless of whether or not the payment is really in full.

Checks are demand paper. Thus, when a drawer's check is presented to the bank, the bank has an obligation to make payment on the check. However, this obligation is removed once the drawer's funds in the bank are exhausted. If the bank fails to meet its payment obligation it is liable to the drawer. However, the payee of the check cannot seek damages from the bank. The bank may refuse to make payment on the check if it is presented to the bank more than six months after the date on the check. In regard to a dispute as to whether a check is still negotiable, the Uniform Commercial Code states that a check should be negotiated within thirty days.

A depositor's bank has a duty to stop payment on a check if informed by the depositor before the check is received by the bank. This practice is commonly followed when a check is lost or a question of payment dispute arises shortly after the drawer writes the check. The drawer's notice to stop the check is referred to as stop payment order. This order can be oral or written. If it is oral it is only good for fourteen days whereas if it is written it can be in effect six months and then renewed. Should the bank mistakenly make payment on a check that has properly been stopped by the drawer, the bank is liable for payment of the check out of its own funds.

One cannot stop payment on a certified check. A certified check is one that the bank certifies by means of setting aside funds for its payment when it is written. The certified check protects the individual to whom the check is addressed (the payee) in that the question as to whether the drawer has sufficient funds in the bank for payment of the check is removed.

Yet another type of check is the cashier's check. The drawer of a cashier's check is the bank. Upon receipt of payment by the depositor, or upon charging the amount against the depositor's account, the bank issues a check on its own funds to the depositor or to the person designated by the depositor. The purpose of the cashier's check is to increase the strength and the ability of the depositor to negotiate the check.

When a person writes a check and does not have sufficient funds in his bank to cover the check, he may be subject to criminal as well as civil action. He is usually ruled to be guilty of fraud. If an individual signs the name of the drawer to a check in order to negotiate the check he is guilty of forgery. The depositor's bank is liable to the depositor for the amount of the check if they make payment on a forged check.

It should be noted that in order to relieve himself of liability in the case of forgery, the drawer must exert reasonable care to prevent his name from being forged. Thus, if a contractor negligently always uses a mechanical writer for signing checks and it comes into the hands of another, he becomes liable for the checks signed with it. In addition, in order to relieve himself of liability of forged checks, the drawer must inform the bank of the forgery within a reasonable time after receiving his bank statement.

If the bank makes payment on a check on which the stated sum of the check has been altered, the bank is liable to the drawer for any amount over the amount stipulated on the check by the drawer. However, as in the case of forgery of signature, the liability of the bank is removed if the drawer contributes to the alteration by means of his negligence.

## Promissory Notes

The second most widely used type of commercial paper by the construction industry is the promissory note. Usually the contractor is the maker of such a note. A promissory note is a written promise by the maker to pay money to the recipient of the note. The recipient, who in effect is a lender of money, is referred to as a payee. Promissory notes differ from other types of commercial paper in that only two parties, a maker and a payee are involved in the paper.

Promissory notes are used when an individual obtains a loan from another party. A contractor often has to sign a promissory note when obtaining loans to finance material, labor, and equipment. To provide the payee (who is often a bank) security, the maker often has to put up property or collateral in the form of stocks or bonds. If property is put up as security, the note is referred to as a mortgage note. If stocks., bonds or another type of collateral is provided, the note is referred to as collateral note.

A promissory note may be either payable on demand of the payee, or it may be payable on a stated date. If the note is payable on a given date, the maker's liability to make payment on the note is not removed if the payee does not present it to him on the due date. The maker's liability continues until it is relieved by the statute of limitations.

A note will often state that it is payable within a certain period of time. For example, it may state that it is payable within 30 days. When such is the case, it is payable within 30 days of the day after the stated date on the note. If the last date due falls on a holiday the note is payable the following business day. If the maker does not make payment by the due date, he is subject to legal action. His action of nonpayment results in the note being dishonored. However, if his delay in paying is caused by circumstances beyond his control, he is granted a delay and must then make payment within a reasonable period of time.

Even when a note is not payable until a stated date, the payee may have the right to accelerate the payment of the note. If the note is such that the maker is to make installment payments on the note, the failure of the maker to meet a payment is grounds for the payee making immediate claim to the entire amount. A note also may be written such that even though a future date is stipulated as the payment date, the payee reserves the right to demand full payment at any time. This is done to protect the payee should it become apparent that the maker is having financial difficulties. For example, a bank may reserve such a right when entering into a promissory note agreement with a contractor. If it becomes apparent to the bank that the contractor is not meeting his financial commitments such as paying his project labor costs, the bank may accelerate its claim on the note.

In regard to acceleration clauses, the maker may also have the right to make early payment. These early payments are referred to as prepayments. They are made to reduce or eliminate interest costs that are associated with the corresponding loan and note. In many cases, the making of prepayments initiates a small charge to the maker. Such charges are not ruled by the courts to be penalty charges and therefore are ruled to be legal.

Like any other type of commercial paper, the payee (the holder) of a promissory note may endorse the note and transfer it. An individual who receives the note from the endorser is referred to as a secondary party. All succeeding parties to which the note is transferred are referred to as secondary parties. If informed of dishonor of payment by the maker, secondary parties can be held liable to the present holder of a note. The holder has a choice of suing any of the former holders to the note. Obviously he will usually make a claim against the party whom he believes is most capable of making payment.

## Drafts

A third type of commercial paper is a draft. A draft is sometimes referred to as a bill of exchange. A draft is a written order of one individual to another demanding an amount of money. The person who issues and signs the draft is referred to as the drawer. In issuing the draft the drawer usually demands that the person to whom he issues it, who is referred to as the drawee or debtor, pay the stated amount to a third party. This third party, who is referred to as the payee is commonly a bank at which the drawer has an account. The purpose of a draft is to require the debtor (the draweee) to pay the money he owes to the creditor (the drawer).

Drafts may be either sight drafts or time drafts. A sight draft is one that is payable on demand. Thus, the drawee has to make payment upon receipt of the draft. A time draft is one that is payable within a stipulated period of time. A certain type of time draft, referred to as a trade acceptance, is commonly used in the construction industry. A construction material supplier will often send a trade acceptance that is payable within thirty days. Besides the legal remedies that the material supplier has upon the contractor's acceptance of the draft, the material supplier can often use the issuance of trade acceptances for the purpose of obtaining loans. He may also actually sell the trade acceptances as accounts receivable to a finance company.

A drawee does not have to accept a draft. However, his non-acceptance of the draft does not eliminate any claim that the drawer has against the drawee. In order to receive material or services, the drawee may have to accept a draft. This acceptance has to be in writing. Upon accepting the draft, the drawee admits to the debt and thus, becomes liable for its payment. The drawee's acceptance of the draft may be either a general acceptance or a draft varying acceptance. A general acceptance is one in which the drawee accepts the draft and does not impose any restrictions on it. On the other hand, a draft varying acceptance is one whereby the drawee imposes a restriction or change on the draft as stated by the drawer. For example, the drawee may stipulate that he will make payment within 60 days rather than 30 days as stipulated on the draft. Naturally, the drawer has the right not to accept the varied condition. However, if he accepts the variation, he is then bound by the changed conditions.

As is true of all commercial paper, the draft can be endorsed and transferred. A drawer might transfer his right to payment from a drawee to another party. If this occurs, the drawer becomes secondarily liable. The drawee is primarily liable to the party who holds the draft.

## 8.7 Employment and Subcontractor Relationships

Besides criminal and tort law, contract law, and commercial paper law, several other areas of law affect the construction industry. Included in these areas are company legal organization law, tax law, and laws pertaining to employment and contractor–subcontractor relationships.

In the normal contractor–owner relationship the contractor is ruled to be an independent contractor rather than an agent of the owner. The legal difference is that an independent contractor is ruled to be acting free from the control of the owner, whereas an agent works under the control of the owner. An agent can make contracts with third parties on behalf of the owner. As a result, the owner can be held for various types of liabilities in an owner–agent relationship that he cannot be held liable for in an owner-independent contractor relationship.

The employee of a contractor may be one of several types. He may be a union laborer in which case the laborer is neither an agent or an independent contractor. Yet another contractor employee may be a job superintendent who has legal authority to be an agent of the contractor. On the other hand, the contractor–subcontractor relationship is often of the type such that the subcontractor is an independent contractor.

Similar to the distinction between an agent and an independent contractor is the distinction between an ordinary laborer or material man and a subcontractor. The Miller Act defines a subcontractor to mean one who performs for and takes from the prime contractor a specific part of the labor or material requirements of the original construction contract.

# 8.8    Employment Law

When an employee of a contractor is a laborer, both parties have traditionally been guided by the various employee laws in determining the benefits that the laborer is to receive. Minimum wage laws require a certain minimum wage. In addition, the Davis Bacon Act requires that labor on a federal construction job is to be paid the prevailing wage rate of the immediate area.

Recent years has seen enactment of several employment acts and programs that affect the relationship between the construction firm and its employees. These employees include union craftsmen.

The most widely publicized of the employment act is the Civil Rights Act of 1964. In effect this act makes it unlawful to fail or refuse to hire or to discharge any individual, or otherwise to discriminate against any individual with respect to his compensation terms, and conditions or privileges of employment because of such individual's race, color, religion, sex, or national origin. In addition it is unlawful to limit, segregate, or classify employees in any way that deprives any individual of employment opportunities because of such individual's race, color, religion, sex, or national origin.

These discrimination laws do not only apply to the construction firm in its employment of labor. In addition it is unlawful for a labor organization, such as a construction craft union, to exclude or expel from its membership, or otherwise to discriminate against, any individual because of race, color, religion, sex, or national origin.

If a court finds an employer, such as a construction firm, guilty of an intentional violation of the Civil Rights Act, it has a wide range of potential remedies including injunction, damages, required hiring, and the awarding of back pay. This last remedy can be particularly burdensome to the employer.

Numerous other employer–employee labor acts are on the books. Included are the Equal Pay Act and the Age Discrimination in Employment Act. Most of these acts are widely publicized to the point where noncompliance is usually intentional. The labor unions have a way of policing the construction firms for possible violation of any employer–employee act.

Construction labor injuries are almost universally covered by workmen's compensation statutes. The first of these state laws was passed in 1913 in Wisconsin, and today every state has workmen's compensation laws. The effect of these laws is that the question of who is negligent, the worker or the employer, is removed in regard to payment for the worker's injuries. That is, compensation for the worker's injury is paid by an independent fund which is financed by employers' payments. A complete discussion of workmen's compensation laws is beyond the scope of this book.

# 8.9   General Contractor–Subcontractor Law

The legal relationship between a project's prime contractor and subcontractors is often complex and may vary from one project to the next. Part of this variation in legal relationship is due to the fact that in some projects the prime contractor selects the subcontractors whereas in other projects the subcontractors may have to meet the approval of the project owner or may be even contracted directly by the owner. The complexity and variability of the contractor–subcontractor relationship makes it difficult to cover the entire legal area in detail. As such only a few somewhat general legal contractor–subcontractor liabilities and relationships are discussed.

The most normal contractor–subcontractor relationship is one in which the prime contractor enters into a contract with a subcontractor to perform part of the work of a construction project. Upon performing his contracted work, the subcontractor is paid the contracted amount of money. As in the case of the payment from the owner to the prime contractor, the subcontractor normally receives payment from the prime contractor in relation to the percentage of his work completed.

Unless the owner directly hires the subcontractor by such a means as separate-but-equal contracts, only the prime contractor is said to be in privity with the owner. Privity of contract is a term that denotes a legal right of contracted duties or subject matter. When a project's subcontractors are hired by the prime contractor, there is no privity of contract between the project owner and the subcontractors. As such, the subcontractor usually cannot sue the owner when the owner causes a delay in the project work. Only the prime contractor (commonly referred to as the general contractor) has legal remedies against the owner if such a delay occurs. Naturally, the subcontractor has privity of contract with the prime contractor when he is hired by the prime contractor.

When the subcontractor contracts with a prime contractor, the contract liabilities of each of the parties are similar to those that exists when a prime contractor contracts with a project owner. However, whereas there is seldom a conflict between the owner and prime contractor as to what work the contractor is to perform, the problem as to who is to do what is a common prime contractor–subcontractor legal dispute.

Typical functions and responsibilities that often are disputed as to whether the subcontractor is to perform are as follows.

1. Who is responsible for various phases of project cleanup.
2. When are progress schedules due and to whom are they to be given (that is, is the subcontractor to make such reports to the owner or to the prime contractor)
3. Who is to provide various project storage facilities.
4. Who is responsible for obtaining permits and licenses.
5. Who is to provide various required safety facilities and programs.

The responsibilities and liabilities of each of the parties should be clearly stated in the contract agreement. A well-written contract is often worth many times the expense of a lawyer's fee for aiding in the writing of the contract. When in fact there is a dispute as to whether the prime contractor or the subcontractor is to perform various work items, the provisions of the contract prevail.

The prime contractor might fail to perform a certain aspect of his project work resulting in the subcontractor not being able to perform his contracted work. When such is the case, the prime contractor is held liable for damages suffered by the subcontractor. In addition, if the subcontractor is delayed in performing his work due to an act of the prime contractor, the courts usually rule that a stated time of completion clause is negated.

In regard to worker injuries, the prime contractor is responsible for keeping the project site in a safe condition for the use of the employees of any subcontractor who are working on the project site. This rule is only in effect when the general contractor is in possession and control of the premises awarded to him by the owner. If he is, then he is liable for the worker's injuries if he is negligent. Obviously, if a worker is injured due to the subcontractors negligence then the subcontractor, not the prime contractor, is liable. If the prime contractor makes workmen's compensation payments for such a worker, the subcontractor is liable to the prime contractor for such payments.

A prime contractor might receive compensation from a subcontractor for allowing the subcontractor to use his equipment on the prime contractor's project. If such is the case, and an employee of the subcontractor is injured while using the equipment as a result of the prime contractor's failure to keep the equipment safe, the prime contractor is liable for the injury. On the other hand, the subcontractor might use the prime contractor's equipment without receiving permission from the prime contractor. When this is the case, the prime contractor is no longer liable for injuries resulting from the contractor's failure to maintain the equipment. The courts rule that the subcontractor using the equipment is a licensee and is not entitled to damages occurred as a result of negligence of the owner of the equipment.

## 8.10   Summary

This chapter has attempted to relate business law to the everyday business of the construction industry. The contractor should not come to the conclusion that upon reading this chapter that he is now a competent lawyer totally capable of handling all his legal affairs. Law is a very large, complex, and changing field of study. Law cases and court judgments concerning the construction industry alone would fill thousands of volumes of books.

However, the fact that the contractor is not capable of handling all of his legal matters does not eliminate the urgency for his knowledge of law fundamentals as they effect his everyday business. Such knowledge can often prevent his involvement in legal disputes and therefore eliminate his need for expensive legal advice and representation. Such legal costs have to be absorbed as an overhead cost and therefore cut into the contractor's profit margin.

Law that affects the non-businessman also affects the contractor. That is, the contractor is not exempt from criminal and tort actions such as negligence, forgery, and fraud.

More frequent in occurrence in the contractor's everyday business are legal disputes arising as a result of his contracts, commercial paper transactions, and relationships with other industry parties such as the subcontractor. A contract must have four legal requirements in order to be valid. It must contain an agreement, competent parties, legal subject matter, and exchange (consideration).

There are several types of commercial paper which the contractor may negotiate. The most widely used commercial paper is a check. Other types include drafts and promissory notes.

The legal relationship between a prime contractor and his employees and project subcontractors varies somewhat from one project to the next. The written contract provides the means by which the relationships of the parties is determined. When a dispute arises, the written documents prevail.

# Chapter 9

# Preparation of the Detailed Estimate and Bid

## 9.1 Components of a Contractor Detailed Estimate/Bid

Perhaps no function is as critical to the success of the contractor as is the preparation of estimates and bids for projects. The majority of construction contracts are awarded to the contractor based on a competitive bidding process. This means that the contractor must prepare his estimate of the cost of doing work, add a desired profit, and then submit the bid to the prospective client. More often than the not, the client will then take the lowest bid as the basis of awarding the contract. On occasion, the contractor will be able to do work on a time and material basis; however this is rare relative to the amount of work that they perform.

The contractor has at least two objectives when preparing an estimate and submitting a bid. For one, given a competitive environment, the firm wants to be the lowest bidder. Secondly, the firm wants to maximize their potential profit. These two objectives somewhat conflict in that in order to be the lowest bid, the firm has to limit the amount profit they put in the bid.

The preparation of an estimate entails several individual steps. Figure 9.1 illustrates the steps involved in the contractor's estimate of his project costs. It should be noted that the estimate fails at the weakest link. If any of the steps shown in Figure 9.1 are performed inadequately, the objective of predicting the real or actual costs that the firm will incur when constructing the project will not be achieved.

When preparing an estimate, the firm will have in hand the project contract documents. These documents consist of the following:

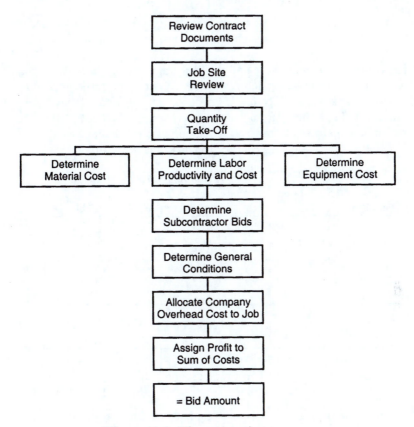

**Figure 9.1.** The estimating process.

- Drawings
- Specifications
- General conditions
- Special conditions (commonly referred to as supplementary conditions)
- Addendum
- Agreement form

The majority of the estimator's time is expended with the drawings and specifications. However, as noted below, other contract documents to include the supplementary or special conditions are also very pertinent to the preparation of an accurate estimate and bid. In addition, a job site review that enables the firm to evaluate job site conditions is critical to the preparation of an accurate estimate.

The time it takes the contractor to perform each of the steps illustrated in Figure 9.1 and the importance of the step in regard to there impact on the final bid amount varies. To this end, the contractor seldom has unlimited time in preparing an estimate for a specific project. The firm should strive to expend estimating time on each step relative to the importance of the step.

The contractor can measure and monitor improvement in the accuracy of their estimates by tracking actual costs of constructing projects relative to the prepared estimate of costs. This post job analysis should only take a few minutes. Using a simple form such as that shown in Figure 9.2, the firm can get the estimator and construction supervisor to review past project performance and critique the accuracy of the prepared estimate.

| Project completed | Estimated cost | Actual cost | [Estimated cost − actual cost]/[a] estimated cost | Reasons for difference |
|---|---|---|---|---|
| | | | | a.<br>b.<br>c. |
| | | | | a.<br>b.<br>c. |
| | | | | a.<br>b.<br>c. |

[a]The absolute value between the estimate and actual cost is calculated.

**Figure 9.2.** Form for post job analysis.

The purpose of this post job analysis is to enable the analysis of differences in what was supposed to happen (i.e., the estimate), and what did happen (actual cost). By getting the estimator and construction supervisor together to review the project results, both individuals should seek improvement.

The accuracy of the overall estimate will only improve if the estimator will address individual components in the overall estimate that are illustrated in Figure 9.1. With this objective, we will not discuss each component individually.

## 9.2   Contract Document Review

The majority of the contractor's estimating time with the contract documents consists of the firm expending many hours taking-off the project work quantities. This is discussed below. In addition, the estimator makes frequent reference to the project specifications with the objective of determining the quality of work that is required and assessing the work effort it will take to meet the requirements.

However, it is also important that the contractor expend time reviewing in detail the project's supplementary conditions. Owing in part to construction claims, disputes, and perhaps contractor financial failures, the project owner, through their designer, is continuing to insert clauses in the supplementary conditions that can have a significant cost implication to the contractor. Some of these clauses are referred to in the industry as "killer clauses". If overlooked, they may lead to a financial loss on the project or even financial failure of the firm.

The number and type of clauses that can have project time or cost implications to the contractor vary from job to job. The best way for the contractor to avoid missing critical clauses is to utilize a checklist such as that shown in Figure 9.3. The checklist shown can be viewed as a "reminder" list; things to look for when reviewing the supplementary conditions.

| Date | Initial | Description | Project cost impact | Project time impact |
|---|---|---|---|---|
| | | Do we get paid for material when installed or when brought to the job site? | | |
| | | Is there a notification of problems clause; how many days is it? | | |
| | | Is there a no damage for delay clause? | | |
| | | Is there a liquidated damage clause? | | |
| | | Is there an arbitration clause? | | |
| | | Do we have to submit project schedules? | | |

**Figure 9.3.** Checklist for supplementary condition.

The purpose of the first two columns is to identify the individual that looks for the detrimental clauses. An individual has to initial each item in the second column, and in the first column indicate the date the item was reviewed or checked.

The above represents only a few of the many contract clauses the contractor should look for when reviewing the supplementary conditions. The firm should add to the list as it reviews additional contracts. The purpose of the checklist or reminder list is to help the firm avoid making the same omission twice.

## 9.3   Job Site Review

The contract documents do not disclose everything about a specific project that may have a project cost or time impact to the contractor. For example, the availability of haul roads, and possible laydown areas for material may not be clearly shown on the contract drawings.

The contractor is not forced to perform a job site review. However, the question is can the firm afford not to do such a review. Undisclosed job site

conditions can have a significant project cost or time impact. Similar to the
suggested contract document review or reminder form shown in Figure 9.3,
the contractor should prepare and utilize a job site review form similar to
the one shown in Figure 9.4. The purpose of the checklist or reminder list
is to help the firm avoid making the same omission twice.

| Date | Initial | Comment | $ Impact | Concern/Description |
|------|---------|---------|----------|---------------------|
|      |         |         |          | Access to job site to include haul roads |
|      |         |         |          | Availability of material suppliers |
|      |         |         |          | Availability of subcontractors needed |
|      |         |         |          | Utility obstructions |
|      |         |         |          | Unusual project obstructions |
|      |         |         |          | Special labor concerns |
|      |         |         |          | Job layout concerns |
|      |         |         |          | Job security |
|      |         |         |          | Unusual site conditions |

**Figure 9.4.** Job site review concerns.

## 9.4    Quantity Take-Off

Perhaps the most basic as well as the most important aspect of the contractor's estimating and bid functions is the taking-off of work quantities from
the two-dimensional drawings representing a project. This is sometimes
referred to as the quantity take-off or quantity surveying.

The take-off process is by far the most time consuming step of the
overall estimating process. The contractor can take-off quantities using one
of three different methods:

1. Manually using a pencil, take-off paper, and a calculator.
2. Using mechanical and electronic aids such as pens attached to a computer.
3. Using output of various computer aided design programs.

The majority of estimators still use the manual process. Given the need
to expedite the take-off process, this will eventually change. Mechanical
devices include roller pens that are moved along the lines and areas depicted
on the contract documents. With the aid computer software; lengths, areas,
and volumes are calculated automatically.

The trend in the construction industry is that the quantity take-off
function will be performed as output of the design function. Computer
aided design software such as AutoCAD have the ability to aid the designer
in preparing the drawings for a project. In addition, these programs have
the capability to output the quantity take-off as a by-product of the design
function. It follows that as time goes on, the contractor will have to expend
less and less time doing the physical and time consuming take-off function;

the computer will do it. To this end, the contractor should ready itself for this transition by taking courses to become competent with computer aided software programs.

For the immediate future, the quantity take-off function remains a time consuming manual process. The contractor has two objectives in performing the take-off function: (1) to do the take-off function accurately, and (2) to perform the take-off function in an expeditious manner. The firm seldom has unlimited time to do the estimating function; especially when one considers the fact that many estimates are prepared as a basis of obtaining even a single project.

The end result is the contractor should use a structured process when doing the take-off function. This entails using shortcuts when appropriate and having multiple estimators do the same process. If the same process is followed every time, the contractor will gain efficiency as well as promote consistency and quality. The following procedures are a few practices that address the accuracy and effective take-off objectives.

- Use a standard order of take-off.
- Use checklists to make sure items are not missed; for example use a material and insurance checklist.
- Use standardized forms for quantity take-off.
- Use preliminary calculations (math calculations that are used several times in the take-off should be made, checked for accuracy, and not done again and again).
- Be consistent when rounding off or converting feet and inches.
- Foot and crossfoot rows and columns of numbers.
- Avoid duplicate calculations.
- Establish a "trail" on the quantity take-off form to the drawings. The reader or checker of the take-off should be able to easily follow the take-off and trace calculations back to the drawings.
- Keep an insight as to the importance and non-importance of specific items when doing the take-off; don't spend 80 percent of your take-off time with items that only comprise 20% of the cost.

The checklist of material noted above serves as a reminder list. By using the reminder list, the estimator can help reduce the possibility that they missed items. Obviously every project is unique leading to the possibility that new materials and items will appear. These should be added to the checklist. The idea is simple, to miss an item once is a mistake. There is no excuse for missing it twice. This is the purpose of a checklist such as the abbreviated one shown in Figure 9.5.

The contractor should use the checklist for each and every take-off project. The time it takes the contractor to prepare an extensive reminder checklist and use the checklist is minimal compared to the negative results that can happen if a single take-off item is missed.

| Date | Initial | Item or material |
|------|---------|------------------|
|      |         | Bulkhead forms with keyways |
|      |         | Edge forms |
|      |         | Screeds |
|      |         | Dowels bars |
|      |         | Supports for dowels |
|      |         | Slabs on grade, direct chute |
|      |         | Slabs on grade, pumped |
|      |         | Slabs on grade, crane and bucket |

**Figure 9.5.** Material take-off reminder checklist.

The estimator can also use various checks on the take-off process as a means of ensuring that all items have been taken-off. For example, there should be a certain expected range of pounds of rebar for each cubic yard of concrete. Similarly there should be a certain range of accepted forming material for each cubic yard of concrete. While each of these items is taken-off individually, the estimator should use rule of thumb checks to validate the accuracy of each item.

## 9.5 Pricing Labor

Labor costs consist of on-site craftsmen costs. These costs are the most variable and unpredictable of all construction costs. They are often the largest and most variable cost element in the estimate. They clearly represent the most risk in the overall construction process.

The determination of the labor cost entails the determination of the following:

- Labor rate (per worker, per crew)
- Productivity of worker and crew doing specific work tasks

While the labor rate is essentially an accounting process and the easier of the two components to determine, the rate must recognize various payroll checkoffs and also forecast any overtime hours worked (the overtime rate is normally one and one-half times the regular rate), and an escalation in the wage rate if the project is projected to overlap a pay period increase.

Labor productivity is the most difficult element of the estimate to forecast with certainty. The labor productivity of a worker or a labor crew is dependent on many variables to include the following:

- Skill of the worker
- Attitude of the worker
- Degree of supervision
- Environmental factors such as temperature and precipitation

- Complexity of work
- Degree of tolerances accepted in inspection process
- Availability of workers
- Job site layout and access
- Cleanliness of project

The above list is only a partial list of the many factors that increase or decrease the productivity of a worker relative to what is estimated. A more exhaustive list of factors and the measurement and improvement of productivity are discussed in another chapter of this book.

Needless to say, the forecasting of labor productivity will never be a science; the contractor will always have risk with this component of the estimate. However, the risk can be reduced and minimized through the collection and use of historical productivity data obtained from past projects. In this regard, the many estimators that are preparing labor cost estimates prepare the labor productivity component using one of the following three approaches.

- Estimate from experience (i.e., using the historical data in the head of the estimator)
- Use of reference manuals such as RS Means Cost Data Books
- Use of structured data base prepared by the firm.

Estimating from experience is still the normal in the construction industry. The estimator, having completed the take-off, essentially looks at the work that is required and based on his familiarly with the work, sets out the number of hours it will take a worker or a crew to do the work. The difficulty with this is that the contractor in effect places total dependence on the estimator making the judgment. Should the estimator error in his or her assumption, or overlook the difficulty of the work to be performed; the result will be an inaccurate productivity forecast. Even worse, if the estimator should leave the employment of the contractor, he or she takes their knowledge with them, leaving the contractor in the dark. Estimating from experience is sometimes referred to as "seat of the pants estimating"; it is not recommended as the primary source for predicting labor productivity. Instead, this approach to estimating should be supported by a formalized data base approach that is discussed below.

Estimators also use published estimating reference manuals. These manuals summarize "average" data that the publishing company has gathered from a variety of sources. An example of one of these is shown in Figure 9.6.

This example data infers that a crew of six workers should be able to place 458 square feet of slab forms a day when using metal pans. Given the six workers on the crew, this equates to .105 man-hours per square foot of forms. The cost data shown reflects an assumption regarding the hourly rate of a worker, material unit costs, etc.

| Description | Crew | Daily output | Man-hours | Unit | Mater-ial | Labor | Equip-ment | Total | Total incl O&P |
|---|---|---|---|---|---|---|---|---|---|
| Form floor slab with 30″ metal pans, 2 uses | C-2[a] | 458 | .105 | S.F. | 1.58 | 2.40 | 0.08 | 4.06 | 5.62 |

[a]Where Crew C-2 consists of 1 carpenter foreman, 4 carpenters, 1 building laborer, 4 power tools.

**Figure 9.6.** Example of historical productivity data.

There are obvious weaknesses associated with the contractor using the above type of data for estimating their own work. The data shown in Figure 9.6 is "average" data. It does not reflect the specific skills of the workers to be used for a specific project, it may not reflect the weather conditions that are forecasted for the project in questions, etc. In other words, it is not job specific. Nonetheless, it is recommended that the contractor have at least one of these reference manuals available to the firm when preparing their own estimate of labor productivity. At the very minimum, they can be used as a check on the accuracy of the firm's own historical data.

The best way to take the uncertainty out of the estimating of labor productivity is through the continual buildup of a data base of past performances. By gathering more and more data about productivity achieved on past projects, the accumulated data can be used to predict the future. While past data will never prove one hundred percent accurate in predicting the future, the science of mathematics indicates that the more data one has about the past, the better the data will be to predict the future; this can be proven with statistics and confidence limits.

Shown in Figure 9.7 is the collection of historical data regarding a single work method (concrete forming measured in square feet of contact area). Similar data can be gathered for each and every work task (commonly referred to as work items) that the contractor performs.

| Past project number | Productivity (mh/100 sfca) | Cumulative avg. (mh/100 sfca) |
|---|---|---|
| 1 | 7.0 | 7.0 |
| 2 | 10.0 | 8.5 |
| 3 | 8.0 | 8.33 |
| 4 | 12.0 | 9.25 |
| 5 | 6.0 | 8.6 |

**Figure 9.7.** Past project productivity data for a forming task.

The data shown in Figure 9.7 represents productivity data for five past projects. After five projects, the data indicates that the average productivity has been 8.6 man-hours per 100 square feet of contact area placed. Admittedly the data shown is somewhat flawed in that it assumes that each project was of the same size in regard to the quantities performed.

If this was not true, the cumulative productivity column entries should be weighted for the amount of quantities placed per project. The point is that by having the above data summarized either on a sheet of paper or a computer screen, the estimator can base his or her prediction for the next project based on the results of prior projects. This will help aid in preparing an accurate estimate. The more data the firm has, the better is should serve the objective of predicting the future.

An improvement on the data base shown in Figure 9.7 is the data base shown in Figure 9.8. In this data base, not only is the average productivity kept, but a measure of the "risk" of the data is calculated. While averages are important, knowing the possible variation of the productivity of the average is equally important in the uncertain world of contractor.

| Past job | Productivity (mh/100 sfca) | Cumulative avg. productivity | Sum of vari- from avg.[a] | Average variance[b] | % Variance = risk[c] |
|---|---|---|---|---|---|
| 1 | 7.0 | 7.0 | 0 | 0 | 0 |
| 2 | 10.0 | 8.5 | 3.0 | 1.5 | 17.6 |
| 3 | 8.0 | 8.33 | 3.33 | 1.11 | 13.3 |
| 4 | 12.0 | 9.25 | 7.0 | 1.75 | 18.9 |
| 5 | 6.0 | 8.6 | 9.6 | 1.92 | 22.3 |

[a]Calculated as cumulative summation of the absolute difference between the average productivity and the productivity for the job sample.

[b]Calculated as the sum of the variation from the average divided by the number of samples.

[c]Calculated as the average variation divided by the average productivity times 100.

**Figure 9.8.** Historical productivity data and risk data for a sample work method.

The last three columns in Figure 9.8 calculate the variation or risk of the samples from the average. For example, after Project #2, the average productivity is 8.5 man-hours per 100 sfca. However the first project varied 1.5 from the average (7 from the 8.5), and the second varied 1.5 from the average (10 from the 8.5). The sum of the two variations is 3 as shown in the fourth column in Figure 9.8. The average variation is 1.5 (3 divided by 2) as shown in the next column. Relative to the average, the variation or risk is 17.6% (1.5 divided by the average of 8.5). This last column is important in measuring the risk associated with various alternative work methods and in evaluating the risk of the overall estimate.

Some estimators argue that the value of historical data is reduced by the fact that every project the firm constructs is different from the past projects; i.e., no two projects are ever alike. In part this is true; however projects and even more importantly, work tasks have similar characteristics. The data base template illustrated in Figure 9.8 is enhanced in Figure 9.9 through the identifying and collection of various factors that influence productivity that were present during the performance of specific projects. Based on the assumption that the productivity of workers is affected by the quality of the supervisor, the contractor can record the name of the specific project supervisor in the column labeled "F1". Other factors such as weather conditions might be summarized in the F2 column. Other "F" columns should be used for additional factors that the contractor judges relevant to having a factor on productivity for the work method in question. By having this data in front of him or her, the estimator can review past samples that is representative of the project being estimated. For example, if the contractor knows that a supervisor named Joe is going to the superintendent on a project being estimated, then the estimator should put more reliance of the past project data samples where Joe was the supervisor.

| Past job | Prod. (sfca) | Cum avg. prod. | Sum of var. from avg.[a] | Avg. var.[b] | % Var. – risk[c] | F1[d] | F2 | F3 | F4 |
|---|---|---|---|---|---|---|---|---|---|
| 1 | 7 | 7 | 0 | 0 | 0 | | | | |
| 2 | 10 | 8.5 | 3 | 1.5 | 17.6 | | | | |
| 3 | 8 | 8.33 | 3.33 | 1.11 | 13.3 | | | | |
| 4 | 12 | 9.25 | 7 | 1.75 | 18.9 | | | | |
| 5 | 6 | 8.6 | 9.6 | 1.92 | 22.3 | | | | |

[a]Calculated as cumulative summation of the absolute difference between the average productivity and the productivity for the job sample.

[b]Calculated as the sum of the variation from the average divided by the number of samples.

[c]Calculated as the average variation divided by the average productivity times 100.

[d]To be used for specifying conditions that prevailed that affected productivity; for example, F1 might be used for the superintendent name, F2 for weather conditions, F3 for location, etc.

**Figure 9.9.** Past project productivity and risk data and productivity factors for a work method.

Given the use of computers, the management of the data shown in Figure 9.9 becomes very easy for the contractor to use. The computer has the ability to sort and search the many data samples that the contractor enters to the template shown in Figure 9.9. It takes only a matter of seconds for the computer to reformulate the data in Figure 9.9 to serve the needs of the estimator.

In regard to computers, the contractor can utilize one of two types of computer program types to accomplish the task of data management. The firm can use what is referred to as "general application programs" like Excel or Lotus to formulate the data base. These programs have templates that can be configured to serve the data base objective. As an alternative, the firm can purchase what is referred to as "industry programs" that have already been custom designed to serve the construction industry's typical needs and format. While industry programs are easier to use, they typically require the user to structure their manual estimating procedures to meet the format of the program. Each approach, the use of a generalized program, or the use of an industry program, has strengths and weaknesses.

Regardless of which approach is taken, the point to be made is that the only good way to reduce the risk of predicting of labor productivity is through the historical data. The more historical data the contractor has, the better the sum of the data will be.

In defining work tasks (often referred to as "work items") the estimator has considerable leeway; no one tells the estimator how to breakdown the overall work. For example, are all pad footings one work item, or should the estimator break down the cost of each size of footing separately? In determining the degree of breakdown, the estimator should consider various criteria to include the following:

- The work should be defined compatible with the concept that if a unique set of resources is required to perform the work, a work item is defined for that work.
- The work items should be defined such that any work that results in a substantially different productivity, even if the same resources are required, should be defined as a separate work item.
- The work items should be defined such that accurate data can be collected at the job to serve the collection of a historical data base objective.
- The work items should be defined compatible with the billing system used to bill the client.
- The work items should be defined compatible with the concept that if a certain percentage of the work is in place (say 50%), and everything is going according to plan, then a corresponding percentage of the work effort (typically measured in labor hours) is expended.
- The work items should be defined compatible with the scheduling and control system of the firm. This enables a more efficient and integrated approach to overall project management.

This last criteria, that of integrated estimating, scheduling, and control is especially important. It is illustrated and emphasized in the project control chapter in this book.

## 9.6   Estimating Material Costs

The cost of materials for a project, while large in total dollar amount, is normally a relatively low risk cost in the overall estimate. There are essentially two elements of estimating material costs for a project:

1. Determination of the quantities of materials for a project.
2. Determination of the cost for the materials.

The determination of the quantities of materials is a by-product of the quantity take-off process. This process was discussed above. Through the take-off process, the type and quantity of required materials are determined. Uses of material checklists aids in an efficient and accurate take-off.

When purchasing the materials and estimating the cost of the materials the estimator must also account for anticipated material wastage. For example, material will be lost in the cutting and fabrication of forming material. The best way of determining the material wastage factor is through keeping historical data from the performance of past projects.

The process of costing the required materials and the risk associated with the estimate depends on the alternative means the contractor has for obtaining needed materials. These include the following:

## Purchase material prior to bidding the project

In this case, the contractor negotiates with an outside supplier that, if the contractor is successful in being the low bidder, the firm will be able to buy the material at an agreed upon price. In this case, the contractor minimizes or eliminates the risk prior to the bid. The material vendor takes the risk of increasing prices. However, the material vendor likely hedges his risk by agreeing to relatively high prices.

## Purchase material after being awarded the contract

In this case, the contractor uses an estimate of material costs when preparing the estimate and "buys the material out" upon being awarded the contract. In effect, the contractor realizes a profit or loss on the material (relative to his estimate) as soon as the material is purchased.

## Purchase material when needed

In this case, the contractor purchases the material when the construction process calls for it. The contractor takes all the risk of material cost escalation until the material is purchased. However, the firm has the advantage that no funds are expended until the material is needed.

## Maintain an inventory of needed material

If the contractor has ongoing need for specific types of materials, he may choose to maintain an inventory of the material; in effect he keeps a "stores account". Material is taken from inventory and assigned to specific projects. If this practice is used, the estimating and costing of the materials is dependent on material costing alternatives the firm has available; for example, first-in-first-out (FIFO), last-in-first-out (LIFO), etc.

To illustrate the differences, consider the following example. A contractor maintains an inventory of materials that he has purchased via the following orders:

| Order # | Quantity purchased | Unit cost ($) | Total inventory value ($) |
|---|---|---|---|
| 1 | 100 | 5.00 | 500.00 |
| 2 | 200 | 5.50 | 1,100.00 |
| 3 | 200 | 6.00 | 1,200.00 |
| 4 | 100 | 6.25 | 625.00 |

Let us assume that in estimating and costing an upcoming project, the contractor needs 300 units of the material noted above. If an inventory method of costing referred to as first-in-first-out (FIFO) is used, the material would be estimated at $1,600. (Note: The first 100 units at $500 from inventory and the remaining 200 units at a cost of $1,100). However, if the last-in-first-out method is used, the same project would be estimated at $1,825. (Note: The last 100 units purchased at $625 and the other 200 units at $1,200.) Obviously the later method would assign a higher cost to the project. LIFO assigns a higher cost to the project in periods of increasing costs. However, using LIFO, the remaining inventory would be valued at a lower cost. The point to be made is that the cost estimate would be different when using the two methods.

## 9.7 Equipment Costs

The contractor is required to use various types of construction equipment to construct a project. If the equipment is rented, the estimating risk mainly relates to how long the equipment will be needed at the project. The hourly rate of the equipment is in effect established by the outside vendor. The risk in the estimated cost of needed rental equipment entails estimating the productivity of the equipment and the working conditions such that the duration each piece of equipment is needed can be determined.

In estimating rental equipment costs, the contractor should prepare a worksheet of all equipment needed on the project being considered. The duration that each piece of equipment is needed should then be estimated and multiplied by the agreed to rental rate. A worksheet such as that shown in Figure 9.10 can be used for this purpose.

The use of owned equipment creates increased estimating concerns and issues. In addition to have is to determine the duration that any specific piece of equipment is required, the contractor must also determine how much it costs to own the equipment on an hourly basis.

In practice, contractors determine the hourly rate of owned equipment in one of two manners.

| Equipment needed | Estimated duration required | Rental rate | Duration times rental rate |
|---|---|---|---|
|  |  |  |  |
|  |  |  |  |
|  |  |  |  |
|  |  |  |  |
|  |  |  |  |
|  |  |  |  |
|  |  |  |  |
| Total |  |  |  |

**Figure 9.10.** Equipment needed form.

- Use of published rental rates.
- Determination of hourly rates through the keeping of their own equipment usage and cost data.

The use of rental rates for owned equipment is not recommended. They do not reflect the actual ownership costs for a specific firm. The actual hourly cost for a specific contractor depends on the number of hours the firm uses the equipment, the wear and tear on the equipment, and how the contractor maintains the equipment.

The hourly cost of ownership of equipment is determined as follows:

Hourly cost = Depreciation

+ Maintenance

+ Operating

+ Repairs

+ Finance cost

+ Insurance cost

+ Property tax

+ Replacement cost

The determination of the above equipment cost components as well as the hours of use are determined from the keeping of past performance data. Accurate and timely equipment record keeping are required to determine this information. This process of keeping historical data to determine the true hourly rate of owning a specific piece of equipment is described in the productivity chapter of this book. This process of establishing and monitoring the hourly rate is not only important for estimating, put is a critical component of overall profitable equipment management.

Once the equipment ownership rates are established, the contractor can use the worksheet shown in Figure 9.10 to prepare the equipment cost estimate for a specific project. The only difference is that the ownership rates are now substituted for the rental rates.

## 9.8  Subcontractor Costs

Owing to the specialty characteristics of certain project work, the contractor may choose to subcontract some of the project work. This work must be included in the project estimate of cost. The contractor does not get directly involved in the preparation of his subcontractor bids. But the firm does have three specific responsibilities in regard to the subcontractor portion of the overall project estimate. They are the following:

1. Solicit subcontractor bids.
2. Evaluate accuracy of subcontractor bids and select subcontractors.
3. Include subcontractor bids in overall project estimate to include the determination of overhead to be applied to subcontractor work.

The competitiveness of the overall contractor's estimate is in part dependent on the competitiveness of the subcontractor bids received. To ensure that subcontracted work is contracted at an equitable and competitive price, the contractor usually requests multiple potential subcontractors (often three or more) to submit a bid for each contract he plans to sub-bid. Assuming each of the competing subcontractors is judged to be qualified to do the work, the contractor awards the contract to whichever one submits the lowest bid.

Perhaps the best means the contractor has for determining the accuracy of subcontractor bids is to compare several competing bids. If the firm obtains only a single bid for a type of work, the firm has little assurance that the bid is equitable. However, if three independent bids are made, all of which are within an expected dollar range, the contractor has a means of comparing the bids and drawing a conclusion regarding their competitiveness. It could happen that all three bids are high and therefore non-competitive. However, this likelihood is less than the chance that a single submitted bid is high.

Once the contractor has determined that specific sub-bids are competitive and selects the bids, he then includes them in his overall project estimate. Often the time period in which the sub-bids are received, reviewed, and summarized in the overall bid is short and therefore the process is relatively hectic. Thus, the firm must be especially conscious of the accuracy of his mathematical calculations and the firm must take care not to miss or duplicate a subcontract bid in the overall project estimate.

In addition to the dollar amounts of the accepted sub-bids that are included in the overall project estimate, the contractor usually adds a cost for administering and coordinating the subcontractors. One might refer to this cost as markup cost related to "managing" the subcontractors. Typically the competitive nature of the construction industry does not permit the contractor to mark up the subcontractor work at the same rate that they mark up his own work.

# 9.9   General Condition Costs (Job Overhead)

Costs directly related to a specific project in general are often referred to as general condition costs, indirect costs, or sometimes as job overhead costs. Examples of job overhead costs include supervision costs, interest costs, and insurance costs. Job overhead costs are considered traceable to a project, but are not easily traceable to any one specific segment or work task of a project.

In practice, many contactors have recognized job overhead-related costs in the project estimate by using an allocation process; for example, the firm might multiply the sum of a project's labor and material cost by 8% to determine the job overhead cost to apply to a specific project.

Estimating general condition costs by means of multiplying a percent times a sum of various other costs is at best a simplistic approach. The process assumes that job overhead costs are the same percentage of the direct costs for each project. This is usually not the case. For example, assuming supervision is handled as a job overhead cost, it is not true that the amount of necessary supervision on projects is always a fixed percentage of various costs. The amount of supervision required often depends on the complexity or uniqueness of the job.

The correct method of recognizing general condition or job overhead types of costs for a project is to estimate the individual general condition or job overhead costs relevant to the specific project and add them for inclusion in the total project estimate. In other words, job overhead costs should not be allocated. They should be handled as a direct cost to a project in estimating the project's costs.

Similar to the estimating of material costs, the contractor should use a checklist or reminder list for taking-off general condition costs. An example of such a checklist is illustrated in Figure 9.11.

Some of the general condition costs are a function of project time only, for example the monthly rental cost associated with a trailer located at the job site. On the other hand some general condition costs are a function of project size and or activity. For example, the surety bond cost for project is primarily a function of the project contract amount. Similarly, other general conditions costs are a function of time and project size or activity. For example, the trailer telephone costs are a function of project time and also the amount of activity going on at the project at any point in time.

| General condition (fob overhead) cost | Estimated cost a function of time[a] | Estimated cost a function of activity |
|---|---|---|
| Superintendent cost | $ | $ |
| Project engineer cost | $ | $ |
| Support staff | $ | $ |
| Trailer cost | $ | |
| Job site telephone | $ | $ |
| Other on-site utility costs | $ | $ |
| Finance costs | $ | $ |
| Bond costs | | $ |
| Public liability cost | | $ |
| Theft insurance | | $ |
| Permits | | $ |
| Total | | |

[a]There is normally a cost component for which there is a '$' symbol shown.

**Figure 9.11.** Worksheet for estimating general condition cost.

It is not being implied that the contractor should or could estimate each and every job site telephone call as a part of estimating the overall project. The point is that some general condition costs are a function of project time and others a function of project activity or project activity and project time. Unless they are analyzed one by one, they cannot be estimated accurately.

The finance cost listed in Figure 9.11 relates to the fact that the contractor typically expends money for paying labor, material, and other job costs before they are able to bill and collect the money from the client. In addition, it is common for the client to withhold a retainer until the end of the project. The end result is that the contractor typically incurs a financing cost while constructing a project. The firm may have to borrow money or may have to use their own investment money. Either way, there is a finance cost associated with constructing the project.

The finance cost associated with doing business is often viewed as being a company overhead cost and as such is included in that section of the estimate. However, the majority of the finance cost that a contractor incurs relates to the cash flow for specific projects. The actual finance cost for a specific project depends on the project duration, the type of costs incurred as a function of time, and the client's promptness of paying the contractor. By taking these conditions into consideration and also by analyzing the project schedule, the estimator can make a meaningful estimate of the finance cost.

While the analysis of each general condition cost takes time, it can be argued that estimate fails at the weakest link. It is as important to accurately estimate job overhead costs as it is to estimate material costs.

# 9.10    Company Overhead or Home Office Cost

Costs that support the constructing of projects, but are not directly related to the project, need to be allocated to a project. These costs are often referred to as general and administrative costs or expenses. They are also commonly referred to simply as overhead costs, or as company or office overhead. Examples of these costs include the following:

| | |
|---|---|
| Principals' salaries | Professional Dues |
| Home office rent | Publications |
| Fees to consultants | Accounting fees |
| Legal fees | Insurances related to home office |
| Travel and entertainment | Postage |
| Home office utilities | Medical expense for officers |
| Office supplies | Advertising |

Two separate questions must be addressed when applying general and administrative (G & A) costs or overhead to specific projects. First, the contractor must decide on the basis for the overhead allocation process. Second, once an overhead basis is chosen, there remains the matter of the mechanics of determining the amount to be applied to a specific project.

Let us first address the mechanics of applying company overhead to a project; it is a relatively simple estimating process. The objective of the allocation or application of general and administrative (G & A) expenses to projects is to apply every dollar of G & A costs, no more or no less, to all projects over a defined period of time. If a firm uses an allocation basis that falls short of applying $50,000 of G & A, and projects completed yield a profit of $30,000, the firm actually incurs a total loss of $20,000. If the firm falls short of applying some of its overhead, the unapplied overhead is referred to as underapplied overhead. Similarly if the firm applies more overhead to projects than it incurs, the overhead is overapplied.

To illustrate the mechanics of overhead allocation, let us assume a contractor allocates its G & A overhead costs to projects on the basis of the sum of direct labor and material costs. Let us assume that for the upcoming year, the firm estimates it annual G & A as follows:

## Step 1: Estimate of Annual Overhead

| | |
|---|---:|
| Last Year's G & A | $270,000 |
| Assume 10% inflation of costs | 27,000 |
| Assume budget for firm growth | 23,000 |
| Estimated G & A for next year | $320,000 |

The next step is to estimate the amount of the basis for the same time period. Let us assume our example firm is to apply G & A based on the estimate of labor and material costs. Based on the firm's projection of volume for the upcoming year, and the assumption that labor and material cost is eighty percent of our example firm's volume, the basis is determined as follows:

## Step 2: Estimate of $ Cost Basis for Allocation

| | |
|---|---:|
| Estimated volume | $4,000,000 |
| Estimate labor and material % | 80% |
| Estimate of labor and material cost basis | $3,200,000 |

The firm's allocation rate is then determined by combining of Steps 1 and 2.

## Step 3: Determine Overhead Allocation Rate

| | |
|---|---:|
| Overhead costs estimated (G & A) | $320,000 = 10% |
| Labor and material estimate | $3,200,000 |

The calculation determines that the firm will apply G & A to individual projects at a rate of $0.10 for every dollar of labor and material cost estimated.

Let us now assume an estimate is being prepared for a specific project. The G & A is applied to the estimate as follows:

## Step 4: Cost to Apply to a Specific Project

| | |
|---|---:|
| Estimated labor and material costs | $500,000 |
| Overhead to Apply: | |
| Labor and material, costs of | $500,000 |
| Times allocation rate of 10% | $50,000 |

If things go according to plan, at the end of the upcoming year, each and every dollar of G & A will be applied to projects. Unfortunately, not everything always going according to plan. For one, the firm may incur more (or perhaps less) G & A than they anticipated and budgeted. Secondly, and probably more likely, the actual amount of the basis may differ from that used to determine the allocation rate. For example, the above firm may not obtain work at the $4M rate estimated.

For our example firm, let us assume that after six months, the firm only has $1.5M of work. Assuming no seasonality, the firm should have had $2M of work (1/2 of the budgeted $4M). If this rate of volume continues, the firm will only do $3M of work instead of the $4M budgeted. The G & A will be underapplied by $80,000. This is because only three fourths of the anticipated volume of work will be attained and therefore only three-fourths of the $320,000 of overhead will be burdened to projects.

Because the amount of the allocation basis can vary from the initial estimate, or over expending of G & A can occur, it is important for the contractor to periodically review the allocation throughout the year; at a minimum on a quarterly basis. For our example firm described above, after the six month review, the firm should either (1) increase the allocation rate, (2) get more aggressive to obtain the needed volume of work, or (3) reduce some of their G & A expense. If these actions are not taken, the firm is destined to have underapplied overhead at year end. Underapplied overhead is like a loss, it must be adjusted from profit that was reported on projects completed during the year.

The above example assumed G & A were applied based on the firm's labor and material cost. However, this is not always the best basis for each and every firm to include the contractor. Consider two projects. Both projects have a labor and material cost estimate of $500,000. Let us assume that one project takes six months to construct and the other project takes a full year. Using our allocation basis of labor and material cost, both projects would be burdened or allocated $50,000 of G & A. However, the project that takes one full year to construct would clearly take more home office support than the one that took six months.

It is important to remember that the purpose of G & A allocation is to charge jobs for the amount of G & A the firm will actually take to support the projects. If the allocation process does not accomplish this, one job will be overestimated and another job may be underestimated.

The best allocation basis varies from firm to firm. For an equipment intense firm, equipment hours should likely be in the allocation basis. If one could pick only two components for the allocation basis, the author would propose that labor hours in the estimate and the planned project duration should be considered. Much of the home office expense relates to supporting self performed on site work labor costs. Secondly, because most G & A expenses are a function of time, the longer a project takes, the more G & A will be required to support the project.

The contractor needs to analyze their own operations to determine the best G & A allocation basis. The objective should be to match the amount of G & A allocated to the project to the amount of overhead it actually takes to support the project in question.

A final note about the estimating of company overhead or G & A costs. Many firms don't estimate G & A separately from the profit determination. Instead the firm combines the G & A and profit into a single percentage referred to as markup. This is not a recommended practice. Company overhead is a cost, profit is a strategy consideration. The company overhead for a project should be determined through the allocation process described above. The profit should be determined using strategy similar to that discussed below.

# 9.11   Profit Strategy

The profit the contractor adds to his bid proposal represents the amount of money in excess of his costs which the firm desires in return for building a project. Disregarding profit, all contractors bidding on a single project may estimate the same cost for building the project. Thus, the profit the contractor adds to estimated project cost may actually represent the difference between his winning or losing the project contract in the competitive bidding procedure. In reality, various contractors bidding on a single project will have different estimates of project cost. This is because of the different structure of cost information, the different construction methods, and the different take-off procedures used by various contractors. However, even in the case of varying contractor project cost estimates, the profit a contractor adds to his bid often determines whether or not he will win the project contract.

Winning the project contract is not the only consideration when one is trying to determine desired profit. The contractor wants to maximize profits or the project rate of return. If the firm bids too high (not the lowest responsible bid), the firm receives no work and must absorb the cost incurred in making the bid proposal. On the other hand, if the contractor bids too low to win the project contract, the firm risks losing money on the project owing to his small profit margin.

In practice, many contractors include profit in their bid by including it as part of markup. Markup is a term often used to represent the sum of company overhead and profit. As was noted in the previous section on company overhead, the contractor should treat company overhead and profit differently; company overhead is a cost, and profit relates to the strategy the contractor takes in trying to be low bid and maximize the potential for profit.

Procedures used by contractors in determining profit to add to the cost estimate are often less than scientific. The contractor may merely arbitrarily add a profit amount based on a hunch. It is suggested that the profit strategy used by the contractor should be more structured and based on sound business principles.

The contractor's desired profit associated with a specific project must consider long-term as well as short-term considerations. Given the competitiveness of the construction business, the long-term profit goals of the firm must recognize the profit margin of the industry at large. The profit margin is typically small given the competitiveness, on the order of 2 to 5%. The profit margin the contractor uses must also recognize long-term growth objectives.

Using the long-term profit margin as a background, the contractor still will likely want to make profit adjustments based on short-term project considerations. Three short-term considerations are the following:

1. Need for work
2. Anticipated risk associated with the project
3. Expected competition

Obviously if the firm needs work, it will reduce it's desired profit margin to enhance the possibility of getting work. For example, let us assume a contractor has a long-term profit objective of 4%. However, given a lack of what the firm considers adequate work, when bidding a specific project, the firm might only include a 2 or 3% profit in their bid.

Project risk should also be taken into consideration when bidding a specific project; the profit added to the bid should reflect this. There are different measures of project risk the firm should consider. For example, if the contractor is bidding a project in a town in which it has not done work and is unacquainted with the labor work rules, the inspection process, etc., the firm will want to increase its preferred profit margin when bidding the project.

Risk can be viewed as uncertainty. Project risk is somewhat of a subjective issue. However by reviewing a list of risk considerations such as that shown in Figure 9.12, the contractor can attempt to rate overall project risk.

If the contractor should get a high sum total on the worksheet, it is suggested that the long-term profit margin should be adjusted upwards. On the other hand, if the sum total of the factors is low, the profit margin may be reduced; i.e., it is a low risk project.

The third short-term profit consideration noted above is expected competition. Given the competitiveness of the contracting business, the amount of profit a firm can include in the bid must reflect the expected number of competitors that are bidding the project. The more competitors that are expected, the less the profit that can be included.

It is possible to mathematically formulate the amount of profit that should be included in the bid as a function of expected competition. The formulated mathematical model is referred to in the literature as *bidding strategy*.

| Risk consideration[a] | |
|---|---|
| Familiar with job location | |
| Familiar with labor | |
| Familiar with project designer | |
| Familiar with all work required | |
| Project duration (longer relates to risk) | |
| Familiar with other contractors | |
| Familiar with project inspectors | |
| Weather variability | |
| Distance from company office | |
| Availability of supervision | |
| Total | |

[a]Rate: 1 = low risk; 10 = high risk.

**Figure 9.12.** Form for evaluating project risk.

To illustrate the concept of bidding strategy, let us assume that a contractor's estimated cost of constructing a project is indeed accurate. Therefore, on a particular project for which the firm submits a bid, he will either obtain his desired profit (assuming the firm wins the bid), or the firm will receive zero profit (assuming the firm does not win the bid). Owing to the possibility of there being two different profits (depending on whether or not the firm wins or losses the contract for the project), it is necessary to define two different types of profit.

A contractor's *immediate profit* on a project is defined as the difference between the firm's bid price for a project and the actual cost of constructing the project. If we let $B$ represent the contractor's bid price on a particular project, and let $C$ equal the contractor's actual cost of constructing the project, then the contractor immediate profit ($IP$) on the project is given by the following formula:

$$IP = B - C$$

If the contractor submits a high bid (a large profit included), the firm's chance for receiving the contract in a competitive bidding environment is very small. As the firm reduces the profit, and therefore the bid, the firm's chance for receiving the contract increases.

If we assign probabilities of receiving the contract to various bids the contractor considers feasible, we can calculate the *expected profit* for the various bids. The expected profit of a particular bid on a proposed project is defined as the immediate profit of the bid for the project multiplied by the probability of the bid winning the contract. In a competitive bidding procedure, winning the contract implies that the bid is the lowest responsible bid. If we let $p$ represent the probability of a particular bid winning the contract for a project, then the expected profit $(EP)$ of the bid is given by the following formula:

$$EP = p(B - C) = p(IP)$$

Assume that a contractor is interested in a certain project, referred to as Project 101. Assume that the firm's estimated cost for the project is equal to the actual cost of the project, which is $20,000. The contractor has a choice of submitting three different bids for the project. These bids and their probabilities of winning the project contract are as follows:

| Bid name | Amount | Probability of winning contract |
|---|---|---|
| B1 | $30,000 | 0.1 |
| B2 | $25,000 | 0.5 |
| B3 | $22,000 | 0.8 |

The probabilities shown are assumed to be estimated from the contractor's evaluation of the chance of being the lowest bidder. Bid B1 has the highest immediate profit ($30,000 − $20,000, or $10,000). However, because of B1's low probability of winning the contract, it may not be the best bid to make to maximize overall profits. The calculation of the various bids' expected profit is as follows:

| Bid name | Probability times immediate profit | Expected profit |
|---|---|---|
| B1 | 0.1(10,000) | $1,000 |
| B2 | 0.5 (5,000) | $2,500 |
| B3 | 0.8 (2,000) | $1,600 |

It is observed that bid B2 has the highest expected profit. Expected profit may be viewed as representing the average profit the contractor can expect to make per project; assuming the firm were to submit the same bid for a large number of similar projects. Expected profit does not represent the actual profit the firm expects to make on a specific project. For the data illustrated above, the contractor would either make a profit of $0 or a profit of $5,000 if B2 were submitted. However over the long range of winning the bid half the time and losing the bid half the time, the contractor's average expected profit would be $2,500.

Because immediate profit does not reflect the probability of a bid winning a contract, expected profit becomes the more informative profit. Therefore, because most contractors have the objective of maximizing total long-term profits, expected profit calculations are more meaningful than immediate profit calculations, and should be used to determine the optimal profit and bid.

The question can be asked, how would the contractor know the probability of various proposed bids being low (i.e., the 0.1 for B1 above, the 0.5 for B2, etc.). The answer to this question is historical data. It is suggested that every time the firm bids against various competitors, the firm should record the bids of their competitors (assuming the information becomes available). After a period of time of collecting the bidding performance of competitors, the firm will be able to establish trends and would be able to approximate probabilities of various bids being the low bid.

In summary, the contractor's determination of profit to include in a bid will always remain somewhat subjective. However, in strategizing as to the profit to include, the need for work, the risk of the work, and the expected number of bidding competitors should all be analyzed. This should lead to a higher rate of bidding successes and higher project profits.

## 9.12    Putting All the Components Together to Form a Bid

The contractor's bid fails at its weakest link. If any of the components is inaccurate or for some reason overlooked in compiling the estimate and/or bid amount, the total bid will be inaccurate. Such an inaccuracy will likely lead to one of two unfavorable events; either the contractor will submit a bid that is too high and therefore be non-competitive; or even worse, the firm may submit a bid that is too low and therefore be awarded the contract to do the construction work but subsequently lose money on the project.

A contractor typically uses a standard set of forms to prepare each of the discussed components of the overall bid. The forms also provide the contractor the means of integrating each of the components into the summarized contractor bid. The specific estimating forms used by a specific contractor vary depending on the type of work he performs, the size of his operation, and to some degree his preferences. The objective remains the same — to submit a competitive bid that enables the firm to realize a profit.

## 9.13    Summary

The construction process is characterized by estimates. The contractor is typically engaged via the competitive bidding process by which the contractor is hired based on a bid. As a basis of submitting a bid, the contractor has to prepare a detailed estimate.

The preparation of the estimate/bid consists of several steps to include the taking off of quantities, and the costing of the labor, material and equipment costs to do the various taken off work items or work tasks. The quantity take-off process should be thorough and systematic. It is a time consuming function.

Costing of on-site labor costs and equipment costs are the most challenging of the estimating steps. The determining of the on-site labor costs entails the determination of the labor wage rate and the labor productivity. The estimating of the labor productivity can be best determined via the collection and historical productivity data. Similarly, the determination of the equipment costs is dependent of establishing equipment hourly costs and equipment productivity. The productivity can be best established through the collection of historical productivity data.

The contractor must also determine the job and company overhead to be included in the estimate. Job overhead is also referred to as general conditions cost; they are overhead costs that are incurred primarily at the job site. Company overhead costs are also referred to as general and administrative expenses or home office costs. The company overhead costs are allocated to the project estimate.

In determining the profit to be included in the bid, the contractor must attempt to maximize their profit while staying competitive. The profit that should be included in the bid is a function of the expected contractor competition, the contractor's need for work, and the risk of the project.

# Chapter 10

# Project Planning/ Scheduling

## 10.1   Introduction

Project planning and scheduling entail the preparation of a formal "road map" of how the overall project will be undertaken. A project is broken into a series of activities by the project planner and sequenced to show the relationships of the various activities. Work durations are assigned to each activity to enable the overall project duration to be determined.

There are several planning and scheduling procedures or algorithms available to the contracting firm, including the bar chart and various network techniques such as critical path method (CPM). However, formal procedures have not been widely used in the construction industry. Historically, contractors have carried project plans "in their heads", not taking the time to prepare rigorous plans and schedules. There is considerable evidence that the failure to do so weighs negatively on the construction industry. Excessive job-site waiting time, duplication of incorrectly placed work, and non-productive time resulting from unnecessary labor, equipment, or material transportation all stem in part from the lack of thorough project planning and scheduling.

Partly as a result of the increased use of computers in the construction industry and partly because of pressures placed on the contractor by the customer or project owner to prepare plans and schedules, firms have been increasing their use of plans and schedules. Most firms that prepare them use them to set out activity milestone dates and to establish the duration of an entire project. Yet a plan and schedule can also be a means of optimally utilizing project resources, including labor.

After presenting the basis of a formal planning and scheduling techniques, CPM, this chapter focuses on using plans and schedules to optimize job-site productivity. Several planning/scheduling techniques will be illustrated.

## 10.2  What is Project Planning and Scheduling?

A formal project plan and schedule can be viewed as having the following elements:

1. Breakdown of a project into a series of individual "work pieces".
2. Determination of work piece durations and required crews.
3. Sequencing of work pieces in an overall project plan.
4. Calculation of project duration as a function of work piece durations and sequencing.
5. Determination of activity "float times" or possible delay times available to a manager to use during unexpected and uncontrollable events.

Two of the more widely used planning and scheduling techniques for construction projects are the bar chart and CPM. The use of these techniques often results in a graphical presentation of a project such as that shown in Figure 10.1.

| Activity | Start | Finish | |
|---|---|---|---|
| Excavate | Sept 1 | Sept 13 | |
| Obtain Sub-base | Sept 1 | Sept 26 | |
| Obtain Pipe | Sept 1 | Sept 26 | |
| Place Pipe | Sept 27 | Sept 29 | |
| Fine Grade | Sept 29 | Oct 6 | |
| Place Sub-base | Oct 6 | Oct 10 | |
| Compact Sub-base | Oct 10 | Oct 13 | |
| Place Concrete | Oct 13 | Oct 17 | |
| Excess Sub-base | Oct 17 | Oct 23 | |
| Backfill | Oct 20 | Oct 29 | |

**Figure 10.1.** Example bar chart.

For larger projects there may be many more work pieces or activities than those shown in this bar chart. Nonetheless, the princples of preparing a plan and schedule remain essentially unchanged.

## 10.3   Why Undertake Project Planning and Scheduling?

One of the most important and detrimental reasons for low construction-industry productivity is the lack of project planning and scheduling. Failure to plan or schedule activities leads to excessive labor and equipment waiting time, delays related to unavailability of materials, lack of subcontractor coordination, and management's inability to react to unexpected events such as poor weather, equipment breakdowns, or material shortages.

A contracting firm would not bid or start a construction project without first preparing an on-paper, detailed cost estimate-in effect, a plan for costs. However, the same firm will often ignore the need to prepare an on-paper plan and schedule. Instead, the firm might argue that a formal plan is subject to too much uncertainty. In reality, the very existence of a plan and schedule can enable a project manager, superintendent, or foreman to effectively react to the many uncertain, unexpected events that characterize the construction process. The more uncertain the production process, the stronger the need to prepare and use a plan and schedule.

The relationship of formal project planning and scheduling and construction productivity is illustrated in Figures 10.2 and 10.3. Figure 10.2 summarizes the findings of an industrial engineering-based study performed at a building construction project (on a sample day). The figure indicates that direct work was being performed only 50% of the time. Undoubtedly, some construction projects are much better in regard to this percentage of productive work — but some are worse.

As Figure 10.2 shows, all jobs include a significant percentage of waiting time, unnecessary traveling time, and time related to redoing work or to poor communication. These time components of an 8-hour day are all non-productive. They take away from the profit objectives of the contractor, lead to negative work attitudes, and have a significant negative impact on productivity.

A plan and schedule that set out material procurement dates will help reduce delays and waiting time associated with material shortages. Similarly, a plan for labor and subcontractor performance can result in a more productive use of available labor crafts and increase subcontractor coordination, attitude, and productivity.

Another way of quantifying the relationship of project planning/scheduling and productivity is illustrated in Figure 10.3. This figure plots a bar-chart type of schedule for a small project that consists of 10 work activities. Each of these activities requires a finite number of resources if the activity is to be performed effectively. Assume we know optimal crew sizes and have designated the crew-size number inside the block beside the activity to which each corresponds, as shown in the figure. We have also

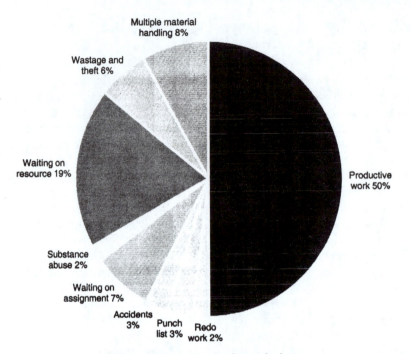

**Figure 10.2.** Example work day.

plotted the number of laborers needed on any one day — maximum of 13. (The actual number needed on any one day varies significantly from day to day.)

It is unlikely that the contractor can hire and fire laborers (or any other resource) on a daily basis according to the labor demand curve shown in the figure. Instead, the contractor probably has a non-optimal number of laborers at the job site on any given day. For example, assume that the contractor keeps 13 laborers at the site every day for the project's duration. This is illustrated by the horizontal line (solid and dashed) at 13 laborers. The areas marked "X" represent "planned non-productive time" a laborer is actually scheduled to be non-productive.

If a superintendent has 5 person-days of work scheduled on a given day and has 10 workers there to perform the work, the 10 workers will produce 5 person-days of work; that is, they will be 50% productive. The challenge of the planning and scheduling effort thus becomes evident. Until a project plan and schedule can be prepared that results in a daily matching of availability of and demand for project resources, there will be "planned" non-productive time.

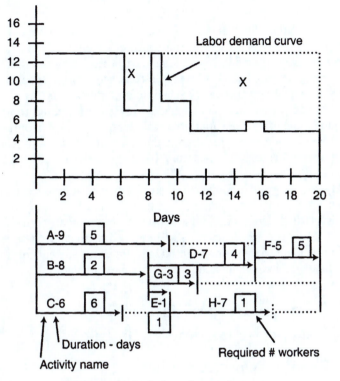

**Figure 10.3.** Labor demand curve.

# 10.4 Implementing a Project Plan and Schedule

A project plan and schedule should be a means of integrating all the entities of a project into an overall team: the contractor's project manager, superintendent, foremen, and subcontractors. The plan and schedule fail at the weakest link.

The difficulty of preparing a plan and schedule is compounded by the organizational structure of the construction project. For example, an in-the-office project manager may have responsibility for putting together a plan and schedule before the contractor even assigns a specific superintendent or foremen to the project. Even after they are selected, a few of them may be negligent in providing input to the scheduling effort, which can weaken it.

The following are guidelines for the preferred timing of each step of the planning process.

1. The project manager prepares a "milestone" or conceptual schedule prior to the project bid as a basis for proceeding with the project

    estimate.

2. With input from field personnel, including the superintendent, and from potential subcontractors, the contractor's project manager prepares a detailed plan and schedule as part of the cost estimate/bid.

3. Upon being awarded the contract, the contractor forces subcontractor input and makes any necessary scheduling modifications.

4. The detailed project plan and schedule are communicated to all field personnel, including the superintendent, foremen, and subcontractors.

5. Daily, field personnel are required to make use of the plan and schedule, including performing "short interval scheduling" (described in later sections).

6. Weekly, job progress is documented and incorporated into the plan and schedule.

7. Weekly, with the help of all field personnel, the plan and schedule are revised. Management decisions and preferred actions are determined, to be implemented by the superintendent, foremen, and subcontractors with the objective of controlling project time and cost.

# 10.5   Steps in Preparing and Using a Detailed Plan and Schedule

## Step 1: Defining Work Activities

The first step in preparing a formal plan and schedule is to define the project work pieces or activities. For example, should "form concrete slab", "place rebar in the slab", and "pour concrete in the slab" be defined as three activities, or should one activity, "place slabs", be defined as the activity? At what level of detail should activities be defined? This is a more critical issue than one might at first think. Too detailed a list of activities will frustrate job-site personnel, whereas too few activities will result in the schedule's being of little benefit as a productivity or control tool. This much can be said: a broader list of activities needs to be defined for a milestone or conceptual schedule than for the detailed schedule that will be used at the job site.

    While it is impossible to set out a specific "best" list of activities for each project, one can establish criteria for defining project activities.

1. The activities should be compatible with the intended purpose and use of the schedule.

2. To the degree feasible, the activities should be compatible with the estimate breakdown.

3. The activities should be compatible with field reporting for cost control.

4. The activities should be compatible with the firm's billing system used for making progress pay requests.

5. Any work function that requires a unique set of resources should be defined as a unique work activity.

More often than not, the use of these criteria will result in the contractor's defining between thirty and a few hundred work activities for the detailed schedule. The actual number will depend on the project size, work complexity, the ability to revise the schedule periodically, and the ability of on-site personnel to use the schedule information.

## Step 2: Determining Activity Durations

There is only one accurate and acceptable way to determine activity durations. The duration of every activity must be determined on the basis of the quantity of work, the crew to be assigned to the work activity, and the estimate of the crew's productivity. Like cost estimating, determining activity durations is subject to uncertainty and contains a degree of risk.

What follows is one example of determining activity duration. In this instance, the activity is erecting forms.

Step 1:   Determine quantity of work   8,000 sfca
Step 2:   Estimate productivity   10 mh/100 sfca
Step 3:   Establish crew size   5 workers
Step 4:   Calculate duration:

$$\frac{(10 \text{ mh})(8{,}000 \text{ sfca})}{(100 \text{ sfca})(5 \text{ mh/hr})(8 \text{ hr/day})} \quad 20 \text{ days}$$

The above calculation needs to be performed for each work activity for the project schedule. Figure 10.9 is a form for making these calculations. If an activity is to be subcontracted, the duration should be estimated or determined from subcontractor input.

---

Job name   _____

Activity   _____

Date   _____

Planner   _____

1. Quantity of work to be performed     _____

2. Possible crew sizes

| Type of crew & # workers | Productivity (units/crew hr) |
|---|---|
| _____ | _____ |
| _____ | _____ |
| _____ | _____ |

3. Crew selected

    _____     _____

4. Activity duration

    Hours = Quantity work/Crew productivity = _____

    Days = Hours/Hours per day (8) = _____

**Figure 10.4.** Activity duration form.

## Step 3: Determining Activity Sequencing or Logic

Project work activities must be sequenced to reflect the actual planned progress of the construction project. This sequencing reflects the technology of construction-it is technically impossible to do certain construction operations or activities until certain other tasks are performed. If a project plan and schedule are to be properly prepared, they must reflect three types of logic sequencing.

1. Technical logic (based on the technology of construction)
2. Resource logic (based on availability of resources)
3. Preference logic (which recognizes project economics)

Resource logic addresses the fact that although it may be possible to perform two work activities at the same time (such as forming the north and east slabs), because of limited resources (say, carpenters) it may not be possible to undertake the two activities together. Preference logic addresses the fact that despite technical ability and the availability of resources, a contractor may decide for economic reasons to do one activity after another.

## Step 4: Adjusting Activity Durations for Contingencies

In an environment of uncertainty (such as poor weather, material shortages, equipment breakdowns), it is unrealistic for a contractor to plan for ideal activity durations. Once the activity sequencing is determined and the project plan is sketched, the preparer of the detailed plan and schedule may decide that it is necessary to make some activity duration adjustments. For example, the initial sketch may indicate that certain concrete work will be performed during months when several rainy days are expected, Seeing this, the preparer may want to add a day or two to the duration of the concrete work to reflect the possibility of rain. A plan that does not include such contingencies is unrealistic, misleading, and may prove detrimental to sound management decisions.

## Step 5: Obtaining Subcontractor Input

Construction entails the coordination of many interdependent contractors. The contractor must obtain timely and accurate subcontractor information including input from specialty contractors. On occasion, this becomes a difficult task. The best way to obtain subcontractor schedule input is to require it contractually. If the subcontractor fails to cooperate, it can be penalized: its retainer can be held, or it can simply not be rehired. A contractual requirement that the subcontractor submit a weekly form, such as that shown in Figure 10.5 can be used to ensure that this input is provided. Showing the subcontractor how its timely and accurate schedule information have prevented problems or increased productivity will help to encourage the subcontractor's cooperation.

| Activities required to complete your work | Code or letter for work activity | Estimated duration | Activity must follow activities (list codes) |
|---|---|---|---|
| | | | |
| | | | |
| | | | |
| | | | |
| | | | |
| | | | |
| | | | |
| | | | |

Sketch work plan below via an activity arrow

**Figure 10.5.** Subcontractor input plan for scheduling.

## Step 6: Drawing the Project Schedule

Once activities have been defined, durations determined, and the sequencing of the activities determined, the project plan can be drawn. This can be done using a bar graph or a CPM diagram. Figure 10.1 is a sample bar graph for a small project. The activities are shown as bars; the beginning of the bar designates the start of the activity and the end of the bar shows the planned completion of the activity. The actual progress of the project is often superimposed on the bar chart alongside the planned progress.

There are several different formats in which to draw the CPM diagram, three of which are shown in Figures 10.6, 10.7, and 10.8. The first diagram uses arrows to represent work activities. Specific planned milestone dates can be set in the circles at the end of the arrows. These milestone dates can be interim progress dates specified by contract or by the project planner.

The alternative CPM shown in Figure 10.7 is commonly referred to as a circle notation CPM diagram or a precedence diagram. In it, work activities are represented by individual circles. The arrows between the circles are used to specify the sequencing or logic of the activities.

In the third alternative CPM diagram shown in Figure 10.8, the individual activity arrows are proportional in length to the activity durations. This CPM diagram, perhaps the most useful as a visual tool, is commonly referred to as a time scale CPM diagram. The dashed lines following the activity arrows represent activity "float" or slack times during which a manager can react to uncertain events. An explanation of the calculation of activity float or slack time follows.

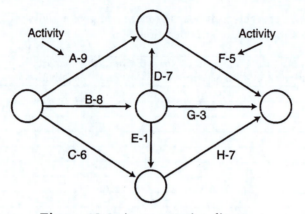

**Figure 10.6.** Arrow notation diagram.

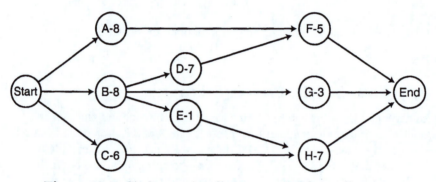

**Figure 10.7.** Circle notation diagram—precedence diagram.

There are thus several different means by which the constructor may draw the overall project plan and schedule. Whatever way is chosen, it is important that the plan and schedule be drawn in a manner that is easily understood by field personnel.

## Step 7: Performing CPM Calculations

Although the establishment of project goals is the major benefit of a project plan and schedule, other benefits can be derived. In particular, if the CPM technique is utilized, calculations can be performed to determine a project's duration, the so-called critical path of its activities, and the float or slack time, which is the management time available for all work activities. These types of information might be called the three objectives of basic CPM calculations. The calculations can be performed efficiently with a computer, but their simplicity makes it possible to perform them without one. Even

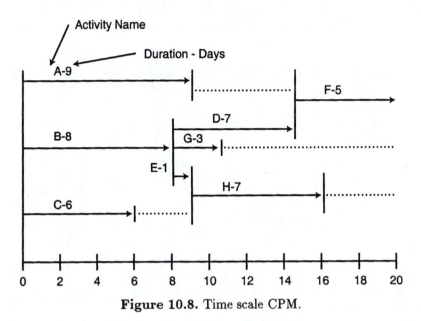

**Figure 10.8.** Time scale CPM.

if a computer is available, knowing how to make the CPM calculations manually enhances the understanding and use of the CPM process.

We can illustrate CPM calculations using the simple CPM diagram shown in Figure 10.6, 10.7, or 10.8 (let us focus on Figure 10.6). The planned duration of each CPM activity is placed beside the corresponding activity. In the CPM diagram shown in Figure 10.6, there is no intended relation between the length of an arrow representing an activity and the duration of the activity.

CPM calculations are made by means of a forward and a backward pass of calculations through the diagram. The forward pass is referred to as the earliest-start-time schedule. In performing the two sets of calculations we will generate five different types of information for each activity:

1. Earliest start time for an activity (EST)
2. Earliest finish time for an activity (EFT)
3. Latest start time for an activity (LST)
4. Latest finish time for an activity (LFT)
5. Total float time of an activity (TF)

The EST for an activity is defined as the earliest possible time at which the activity can start. Assuming a start date for the project as the end of Day 0 (equivalent to the beginning of Day 1), the earliest start times of Activities A, B, and C are 0. (CPM activity durations can be given in hours, days, or weeks; also, by using a calendar we can make a calendar-

day calculations rather than working-day calculations.) The EST answers
along with answers to the other calculations are given in Figure 10.9.

| Activity | Duration | EST | EFT | LST | LFT | TF |
|----------|----------|-----|-----|-----|-----|-----|
| A | 9 | 0 | 9 | 6 | 15 | 6 |
| B | 8 | 0 | 8 | 0 | 8 | 0 |
| C | 6 | 0 | 6 | 7 | 13 | 7 |
| D | 7 | 8 | 15 | 8 | 15 | 0 |
| E | 1 | 8 | 9 | 12 | 13 | 4 |
| F | 5 | 15 | 20 | 15 | 20 | 0 |
| G | 3 | 8 | 11 | 17 | 20 | 9 |
| H | 7 | 9 | 16 | 13 | 20 | 4 |

**Figure 10.9.** CPM calculations.

The EFT of an activity is defined and calculated as an activity's earliest
start time plus the activity's duration. In other words, if Activity A can
start on the end of Day 0 and takes nine days to complete, the soonest A
can be completed is the end of Day 9.

The EST of any given activity is calculated as the maximum EFTs
of activities immediately preceding it. In the case of Activity D, only one
Activity, B directly precedes it. Therefore, the EST of Activity D is equal
to the EFT of Activity B, or the end of the eighth day.

Other activity ESTs and EFTs can be calculated in a similar manner.
Consider Activity F, shown in Figure 10.6. Its start is constrained by Activ-
ities A and D. The EFTs of Activities A and D are 9 and 15, respectively.
Because Activity F cannot start until both Activity A and D are complete,
Activity F's EST is the larger of the two EFTs, or 15. The ESTs and EFTs
of all the activities shown in Figure 10.6 are given in Figure 10.9.

Because Activities F, G, and H terminate the project, the largest EFT
of these three activities is the minimum project completion date. The EFTs
of Activities F, G, and H are 20, 11, and 16, respectively; therefore, the
minimum project completion time is 20 days.

The LST and LFT for each activity are calculated from the backward
pass through the CPM network. As a starting point for the calculations,
the project duration is set equal to the minimum project duration (20 days
in the example in Figure 10.6 and Figure 10.9). The LFT of an activity is
defined as the latest possible time at which an activity can finish without
delaying the predetermined project completion date. Because Activities F,
G, and H are terminating activities in the CPM diagram, their LFTs are
20 (see Figure 10.9).

The LST of an activity is the activity's LFT minus the activity's duration. Because Activity F has a duration of five days and must be completed

by the end of the twentieth day, it must be started no later than the end of the fifteenth day ($20 - 5 = 15$).

The LFT of a given activity is equal to the minimum LST of the activities immediately following it. Therefore, the LFTs of Activities A and D are 15 (the LST of Activity F).

The LSTs and LFTs of the remaining activities shown in Figure 10.6 are calculated in a similar manner. Activity B, for instance, has three activities immediately following it. Thus, Activity B's LFT is the smallest of the LSTs of Activities D, E, or G, which are 8, 12, or 17, respectively. The LFT of Activity B is therefore 8.

The last CPM calculation involves determining each activity's total float time, which is the amount of time for which an activity can be delayed, assuming no other activity is delayed, without affecting the minimum completion date of the project. Total float time is a function of the earliest and latest start times. Total float time for an activity is the difference between the LST and the EST for an activity. It can also be determined by taking the difference between an activity's LFT and EFT. Either of these calculations will yield the same total float value. The calculated total float times for the activities are given in Figure 10.9 for the CPM diagram shown in Figure 10.6. Activities B, D, and F have total float times of 0. The LSTs and ESTs for these activities are equal, as are their LFTs and EFTs. In other words, these activities cannot be delayed. They dictate the minimum project duration and form a critical path through the CPM network. The second basic CPM objective is thus complete, and the critical path is indicated by the activities that have the least amount of total float time (in this case, 0).

Our third objective, that of determining possible float times for the activities, is also complete. The total float times given in the table show the amount of possible time each of the non-critical activities can be delayed without affect the project completion date.

## Step 8: Undertaking Short-Interval Scheduling

The planning and scheduling discussed to this point pertains to the project at large. However, there is another level of planning and scheduling that should take place and can yield equal productivity benefits. This level of planning — short-interval scheduling — entails planning tomorrow's work today.

All too often the project superintendent or foreman reacts to problems at the site rather than taking steps earlier to prevent the problem from occurring. As much as 10% of a day at a site is non-productive because of the need to look for tools or equipment while a craftsperson or entire crew is waiting for them. Not planning tomorrow's work today leads to numerous unnecessary double-lifting and movement of materials.

A simple form such as the one shown in Figure 10.10, if used on a daily basis by the superintendent or foreman in conjunction with the overall project plan and schedule, has the potential to increase job productivity by 10% or more. The form requires that the superintendent or foreman do the following daily (near the end of the day):

1. Set out the type of work to be performed the next day.
2. Set out a quantity goal to be achieved the next day.
3. Set out the tools, equipment, and resources, including labor that will be needed to achieve the work goal.

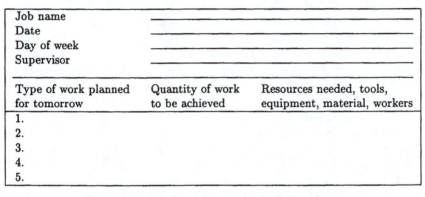

| Job name | _____ |
| Date | _____ |
| Day of week | _____ |
| Supervisor | _____ |

| Type of work planned for tomorrow | Quantity of work to be achieved | Resources needed, tools, equipment, material, workers |
|---|---|---|
| 1. | | |
| 2. | | |
| 3. | | |
| 4. | | |
| 5. | | |

**Figure 10.10.** Short interval scheduling form.

## Step 9: Using the CPM Results

The CPM process can yield benefits over and above the three basic objectives of establishing the latest project completion date, the critical path, and the float or slack times for all activities. Several of the extended benefits, including cash management, good billing and pay request procedures, and proper allocation of resources (such as labor) to the project are best obtained using a computer and accompanying software. Even without a computer, however, relatively simple but practical benefits can be obtained if the project planner is aware of float times. Consider the CPM diagram shown in Figure 10.8. It is a time scale schedule for the arrow notation CPM diagram in Figure 10.6. Suppose that it rains an unexpected amount on Day 2 and 3 and that Activities A, B, and C cannot proceed if it rains. Assume further that in preparing the construction plan and schedule, the construction planner did not anticipate rain. The obvious question is whether the project will take more than 20 days or units of time. At first glance, it may seem so because Activity B, which is on the critical path, has been delayed 2 days. However, Activities A and C have total float times of 6 and 7 days, respectively. Therefore, it may be possible to take resources from these two activities on Days 4, 5, and 6 if necessary and assign them to Activity B to enable it to "catch-up" to its original 8-day duration and maintain a 20-day schedule.

This example is a simple but practical application of the CPM technique. Other practical applications of the concept of float times include using float times to prepare a schedule that better utilizes resources from the point of view of productivity. Figure 10.3 adds a resource requirement (number of laborers required) to each of the activities in Figure 10.8 and also shows the cumulative number of laborers needed on any one day to complete the schedule as shown (the earliest-start-time schedule).

Earlier, we characterized this schedule as one that required the contractor to hire and fire as needed or to hire more laborers than needed on several days. Neither alternative is preferred. Can a better schedule be prepared for the 20-day project that satisfies the technical, resource, and preference logic and makes more productive use of assigned laborers? The answer is yes. It is possible to shift activities within their float times to yield the schedule illustrated in Figure 10.11. This schedule requires only 10 laborers, not the 13 previously required (a 30% decrease). The schedule makes better use of assigned laborers and results in a better matching of availability and demand for resources as well as improved productivity.

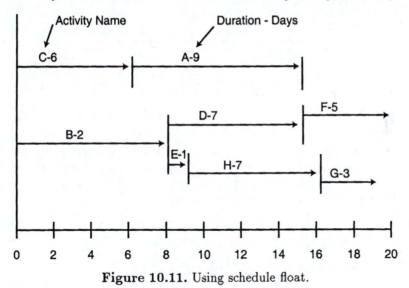

**Figure 10.11.** Using schedule float.

By using the floats that were calculated in Figure 10.9, the scheduler can resequence activities to level the number of workers needed on any specific day. This is just one of the many applications of using float to perform resource management. There are many algorithms that can be used to schedule resources to meet a specific objective. The point to be made is that the floats that were calculated in Figure 10.9 represent available time to the scheduler to meet specific objectives; be it leveling resources, recovering lost project time, or another defined objective.

## Step 10: Updating a Project Plan and Schedule

It is critical that the initial project plan and schedule for a project, be it a bar chart or CPM diagram, be updated as the job progresses. If the initial plan and schedule are not updated, they can lead to poor daily management decisions.

When a 10-month project is 2 months old, one can argue that there is a new 8-month project to be started. The events that have occurred during the first 2 months of the project should be recognized; this recognition may change the critical path, the project duration, or even require adjustments in the duration of or crew sizes needed for remaining activities.

Fundamental to the updating of a schedule to reflect what has happened to date is the obtaining of timely field data. This requires a commitment to accurate field reporting. The use of a daily report such as that shown in Figure 10.12 can be used for this purpose.

| Activities worked on this week | Percentage completed | Need to revise activity durations | Revised duration to complete | Remarks |
|---|---|---|---|---|
| 1. | | | | |
| 2. | | | | |
| 3. | | | | |
| 4. | | | | |
| 5. | | | | |
| 6. | | | | |
| 7. | | | | |
| 8. | | | | |
| 9. | | | | |
| 10. | | | | |

**Figure 10.12.** Form for updating work progress or difficulties.

Actual revisions of the project plan and schedule, including revised CPM calculations, are best achieved with a computer; however, even absent a computer, the need for updating exists, and a project plan and schedule can be updated manually in a simplified form.

## 10.6   Computers and Project Planning/Scheduling

The use of computers, especially microcomputers, has increased in the construction industry. Contractors are now computerizing their accounting and project management functions. Considerable computer software has been developed for constructors, including project planning and scheduling software. These computer programs have the potential to aid project planning and scheduling and therefore to increase productivity.

A computer is not, however, a necessary requirement for formal project planning and scheduling. It is possible to prepare a plan and schedule by hand — one that yields significant productivity benefits. Although the computer can aid planning/scheduling efforts, there are many misconceptions regarding what a computer can and cannot do. By itself, a computer cannot prepare a plan and schedule. It can speed up the process and enable more sophisticated management applications, such as CPM resource applications.

## 10.7  Summary

Constructing a project without a formalized written project plan is like traveling to a distant location without a roadmap. Clearly, if one is traveling to a new location for the first time, and one forgets to take a road map, the time and cost of the journey will be larger than if one took a road map. This is the reason why the contractor needs to prepare a formalized plan for constructing a project. The estimate is the plan for cost; the schedule is the plan for time. Both are essential.

The most common means for preparing a project plan is the use of the critical path method (CPM). A CPM schedule can be prepared with or without a computer software program. By relatively simple calculations, it is possible to determine the earliest start time, earliest finish time, latest start time, latest finish time, and total float for each project activity. This enables the contractor to determine the project's minimum project duration, the project's critical path, and the available float time for each project activity. The available float enables the contractor to resequence activities or delay activities to meet certain project objectives such as recovering lost time, or leveling required project resources.

It can be argued that the preparation and understanding of project planning and scheduling is an essential element of effective project management. A project plan and schedule sets out project time goals and provides a means of project monitoring.

# Chapter 11

# Productivity Management and Improvement

## 11.1  Introduction

Productivity, or the lack of it, is perhaps the number one problem confronting the construction industry, the construction firm, and the construction project. In this chapter, we will discuss how the contractor can measure productivity and the ability or opportunity to increase productivity.

Productivity can be viewed as the efficiency by which materials are placed by labor and equipment. Productivity is commonly measured by means of the following definition.

$$\text{Productivity} = \frac{\text{Units or Dollars of Output (adjusted for inflation)}}{\text{Manhours of Input or Effort}}$$

It should be noted that while this is a widely accepted definition, it can be misleading in that manhours is in the denominator. This might lead one to believe that the only way to increase productivity is to work harder, to make more labor effort. We will illustrate in this chapter that there are many ways to increase construction productivity that do not center on working harder. We will focus on increasing productivity via working smarter and by improved management practices.

A second point should be made regarding the above definition of productivity. Given the definition, productivity for various construction work tasks may be given as cubic yards of concrete placed per manhour (or person hour), board feet of lumber placed per manhour, etc. However, it should be noted that individuals in the construction industry, and in particular the estimator, normally speak of manhours per unit instead of units per manhour. As such, it is more common for the estimator to be citing the reciprocal of productivity.

## 11.2   Low Construction Industry Productivity

Given the above definition of productivity, the United States Department of Commerce has measured the average annual increases in construction industry productivity to be less than one percent a year for the last ten years. The average annual increase in construction productivity of 0.8 percent compares to a 2 to 3 percent annual increase for all U.S. industries.

Unfortunately during the same time period when construction productivity has been nearly flat, the contractor's construction costs have risen. During the past ten to fifteen years, construction costs have increased each and every year, sometimes in excess of five percent in a given year. This increasing costs and relatively flat productivity has put downward pressure on the profitability of many contracting firms. Given the fierce competition of the bidding process, the contracting firm may not be able to pass on added material and labor costs to the customer. For many contracting firms this decreasing profitability owing to the failure to increase productivity while costs have increased, has resulted in risk of the bid exceeding the planned profitability in the bid. This is illustrated in Figure 11.1.

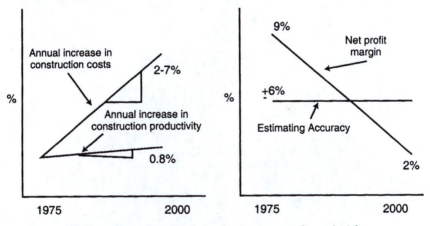

**Figure 11.1.** Productivity increases, profit and risk.

Accounting and estimating studies performed by the author have indicated that on the average, a contractor brings in a project for a six percent different cost than estimated independent of change orders. This is illustrated in Figure 11.1. However increasing costs and relatively small productivity increases has resulted in many contracting firms having their profit margins (after company overhead) decrease to approximately two percent. This fact is evidenced by published financial ratio publications.

It is noteworthy to mention that inflation of costs can be addressed two ways by the contractor: (1) by holding down costs, or (2) by increasing productivity. The construction industry has not done well in either of these and as such one can argue that the cost of construction relative to the value of work put in place has decreased significantly.

Another way of looking at productivity in the construction industry is to look at the composition of the eight hour work day. Documented job site studies performed by the author indicate that between forty and sixty percent of a typical construction day is for non-productive time. One can consider non-productive time to include time associated with workers waiting for instructions, doing redo work, taking advantage of a lack of proper supervision, etc. In addition, non-productive time includes a certain amount of what can be referred to as unnecessary support time such as a worker carrying boards from one location to another merely because the material was not effectively stored in the proper location in the job site layout process. An example of such a work day is illustrated in Figure 11.2. Especially noteworthy is the fact that the same studies performed by the author indicate that approximately one third of non-productive time can be traced to management actions (or lack of actions).

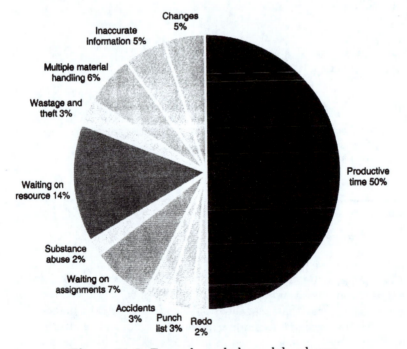

**Figure 11.2.** Example work day—labor hours.

Inspection of Figure 11.2 indicates non-productive time that relates to causes such as poor communications; waiting for material, labor, or equipment; late starts and early quits, etc. These causes of non-productive time can not be blamed on labor attitudes or labor work rules. Simply put, they are the fault of management and can be corrected by improved job site management. Hundreds of thousands of dollars if not millions of dollars of work are performed each year that can be traced to poor communications. The supervisor tells the worker what to do, and the worker may do the work incorrectly owing to the fact that what the supervisor said is not what the worker heard. In addition, the fact that the worker may always be told what to do rather than asked for ideas can lead to a worker attitude that may prove counter productive.

## 11.3   Opportunity to Increase Productivity

It is not being suggested that the solution to the problem of low construction industry productivity or non-productive time is an easy problem to address. The construction process is a difficult one. Problems such as a variable environment to include precipitation, temperature variations, and the complexity of the building process itself are just a few of the problems that most non-construction industries do not have to confront.

Independent of the difficulties associated with improving construction productivity, it should be pointed out that tremendous opportunity exists to improve productivity. If the construction process can be correctly identified as having fifty percent non-productive time, it can also be rephrased as having an opportunity to increase productivity by fifty percent.

It is unrealistic to believe that the contracting supervisor can eliminate all construction non-productive time. However, a mere small increase in productivity on the order of five percent can have a significant impact on the profitability of the construction firm. Shown in Figure 11.3 is a breakdown of a typical construction bid for a building construction project.

A five percent increase in productivity would have the effect of decreasing the overall project labor costs by five percent. As illustrated in Figure 11.3, a five percent increase in productivity and a corresponding five percent decrease in labor costs would result in a profit contribution equal to the initial planned profit. The end result is that a mere five percent increase in productivity can have the result of doubling the profits of the contracting firm. Naturally it also follows that a five percent decrease in actual productivity versus planned productivity, would eliminate any planned profits in the bid shown in Figure 11.3.

Another way of looking at the opportunity of improving productivity and the positive impact of such an effort is to look at the potential impact of the construction supervisor. Previously it was noted that on many projects,

| Bid cost component | Cost % | Example 1M project |
|---|---|---|
| Direct Labor Cost[a] | 40 | $400,000 |
| Direct Material Cost | 35 | 350,000 |
| Construction Equipment Cost | 7 | 70,000 |
| General Conditions — Job Overhead | 8 | 80,000 |
| Company Overhead Cost | 8 | 80,000 |
| Net Profit[b] | 2 | 20,000 |
| Total | 100 | 1,000,000 |

[a]Assume a 5% increase in labor productivity. The result: Labor cost savings = (5%) ($400,000) = $20,000. A 5% increase doubles the "planned" profits.

[b]Assume a 5% decrease in labor productivity. The result: Labor cost increase = (5%) ($400,000) = $20,000. A 5% decrease eliminates the "planned" profits.

**Figure 11.3.** Potential impact of productivity — example project.

one third of the non-productive time at the job site can be traced to the lack of on-site management actions. Given this assumption, and given a typical cost breakdown for a one million dollar project as illustrated in Figure 11.4, one can suggest that one third of two hundred thousand dollars or $66,667 can be traced to poor management or a lack of management.

Assuming that the construction supervisor can implement effective management practices that will eliminate the non-productive time related to management practices, it follows that the effective supervisor can enable an additional $66,667 profits, a number that is in excess of three times the initially planned profit in the bid estimate. There are few industries in which a supervisor can make such an impact on the profitability of the firm. Clearly a supervisor can make a big difference in the construction process!

It should be pointed out that in Figures 11.3 and 11.4 that the benefits of improved productivity may actually be somewhat higher than those calculated. If productivity is improved, the project duration is likely to be decreased. Given the fact that job overhead costs such as trailer rent and supervision costs are almost totally proportional to project duration, it follows that an increase in productivity would also lessen job overhead costs. This would result in additional contributions to project profits. The end result is that a small increase in job site productivity can result in a significant increase in job site profitability.

## 11.4 A Ten-Point Program for Improving Labor Productivity

The measure of on-site labor productivity can make the difference between a profitable and non-profitable project for the contractor. Obviously there

Premise 1: Example Cost Breakdown of $1,000,000 Project

| | |
|---|---:|
| Direct Labor Cost | $400,000 |
| Direct Material Cost | 350,000 |
| Direct Equipment Cost | 70,000 |
| General Conditions (Job Overhead) | 80,000 |
| Company Overhead | 80,000 |
| Profit | 20,000 |
| Total | $1,000,000 |

Premise 2: Typical Non-Productive Labor Cost in Bid

| | |
|---|---:|
| Productive Time (50%) | $200,000 |
| Non-Productive Time (50%) | $200,000 |

Premise 3: Reasons for Non-Productive Time

| | |
|---|---:|
| Labor related (1/3) | $66,667 |
| Industry related (1/3) | $66,667 |
| Management decisions related (1/3) | $66,667 |

Premise 4: Possible Impact of Good Supervisor

Assume an effective supervisor can eliminate non-productive time related to management actions or inactions

Result

| | |
|---|---:|
| Decrease in labor cost | $66,667 |

Possible return of an effective supervisor

| | |
|---|---:|
| Decrease in labor cost/Profit in bid | $66,667/$20,000 |

Conclusion

An effective supervisor can more than "triple" the planned job profits.

**Figure 11.4.** Possible impact of effective supervisor.

is no quick fix to obtaining vast increases in productivity. However, by taking specific proactive actions, improvement can be made. A ten-point program for improving on-site project labor productivity follows:

# Making the project look like a firm rather than a job: Developing a personnel management program.

The construction process is very dependent on the efforts of the construction workers. Typically on projects, the direct labor cost component may be in excess of 35 to 40 percent of the total project cost (the other cost components being material, equipment, job overhead, and company overhead).

The high dependence of labor efforts is compounded by the fact that many construction workers may view themselves as working for a job rather than a firm. Whereas a tool and die worker in a factory, a retail clerk at a merchandiser, or a receptionist at office may work for a firm their entire working life, a construction craftsman may work for several contractors in a given year. One might argue that the construction worker doesn't view himself as working for firms; instead he may view himself working on jobs.

The end result is that the contracting firm and supervisor are very dependent on the work attitude of the construction worker. A worker may have an attitude that he helps the contracting firm by being productive when he doesn't have to, or he may have an attitude that he won't be productive unless he has to. For example, if a worker completes work assigned to him, does he go looking for more work to do, or does he stand idle waiting for more work to be assigned to him? The difference between his attitude is in great part the difference between a productive job and a profitable job and a non-productive low profit or loss of profits job. If the worker views himself as working for a firm he is more likely to be productive when he doesn't have to versus if he views himself working for a job.

The fact that construction workers may view themselves as working for a job versus a firm compounds the personnel management efforts of the supervisor. In his attempt to develop positive worker attitudes and align the worker to the construction firm as well as the job, the author would propose that the supervisor pay attention to the following three needs of each and every worker.

1. Pride in work
2. Measuring system of performance
3. An effective communication channel

Pride in work in great part includes recognition and giving the workers a sense of accomplishment. Personnel management actions such as a mere pat on the back, placing the names of workers on a sign at the job site, asking workers for suggestions and work ideas, and having a day whereby the worker is able to bring his children or friends to view the job site can all be actions in a long term commitment to productivity improvement. While some of these and other personnel management actions are constrained by the short term nature of some jobs and by other constraints to include insurance requirements, failure to show an effort to given workers pride in work is likely to yield negative results.

An effective measuring system entails giving a worker a basis of measuring his own individual performance. This includes communicating what is expected of him, and communicating how he is doing relative to the plan. The plan and subsequent performance system should be communicated both at a job level and at an individual worker level. Leaving the worker in the dark as to what is expected of him and how the project is to progress and how it is proceeding does not accommodate a positive worker attitude. Consider the case of a coach of a basketball team. What would happen if the coach told three of the players to go to the locker room while he explained the game plan to two players. Clearly this would cause discontent, a divided team, and a less than positive attitude for three of the players. Isn't that what a supervisor is doing when he keeps the construction workers in the dark?

The author is suggesting that the supervisor give some thought to sharing information such as manhour budgets and expected productivity for specific work tasks, projected project schedules, and project progress with the workers. The alternative is to assume that the workers don't care. This negative assumption promotes a we versus they attitude that is sure to result in less than satisfactory productivity.

The providing of workers with a communication channel is discussed below as a separate step in developing a productivity improvement program. It should be noted in conclusion that anything the supervisor can do to provide each and every worker pride in work, a measuring system of performance, and a communication channel can be part of an overall effective personnel management program.

## Improved Communications

Poor communications at job sites leads to unnecessary redo work, poor worker attitudes, and an inability to properly monitor the work progress. There are two types of communication that are critical to a productive job site; oral communications and written communications.

Effective and productive communications at a job site is complicated by the fact that communications are carried out in the open and at a relatively noisy job process, and by the fact that the individuals communicating may have different vocabularies and different communicating skills. A construction craftsman, foreman, superintendent, project manager, and architect may all speak and interpret various phrases and words differently.

Effective communications entails listening as well as talking. All too often the contractor supervisor only talks at the worker instead of asking the worker for ideas or listening to his concerns. On occasion the person who knows best how to form the concrete or place rebar may not be the supervisor but instead the craftsman. Failure to take advantage of the workers knowledge runs the risk of not only taking advantage of an improved construction method but also may adversely affect the work attitude of the craftsman. Knowing a better way to do something but not being asked one's ideas tends to promote an "I don't care attitude".

Effective supervisor communications also entails taking the time to properly explain the work process to the worker. The construction craftsman may think he is suppose to know how to do something that is told to him even if he doesn't. Given confusion as to what to do, rather than ask for an explanation, the worker may proceed to do the work incorrectly. The end result is that the contractor will have to correct the work later; a non-productive work process.

The construction industry has been characterized for many years as an industry with inadequate written communications at the job site. Inaccurate time cards, late reports, failure to give the worker or supervisor written feedback, and lost or misplaced documents are typical of the construction job site. Part of the reason for these written communication inadequacies relates to the decentralized nature of the work process. Unlike most industries that create and monitor their written communication system to include their cost accounting process at the same place they make their product, the construction industry is such that written communication is often created at the job site, transferred to the contractor's main office, and hopefully communicated back to the job site. This process results in untimely and sometimes incorrect reports.

The supervisor often complains about bad record keeping they get at the job site; but in fact he himself may promote bad record keeping. The use of a weekly timecard that requires foremen to keep track of workers' hours charged to specific work tasks is likely to be filled out weekly rather than daily. The result will be that the foreman cannot remember on Friday what the worker did on Monday. Perhaps a daily time card would ensure more accurate data. Preprinting time cards with work codes may improve proper charging of labor hours. In addition, the use of daily report forms that require supervisors merely to check items like the weather conditions rather than describe the weather conditions is more likely to be legible and lessen the time for the recording process.

In critiquing their own written communication process or system the contractor should remember the following three rules for improving the accuracy and timeliness of the job site record keeping process.

1. An individual that is required to fill out a form should be showed where the data goes and how it is used.
2. An individual that is required to fill out a form should be shown by example that their data was in fact used.
3. Any individual that fills out a form or inputs data should be given a subsequent feedback report or data.

Given these rules, consideration might be given to posting at a very visible worker location a sign that flowcharts the information system being used at the job site. Also the posting of a sign or report that charts project progress against planned progress such that the workers can measure their progress can also aid communications as well as aid in making the job look like a firm. In summary, the supervisor should work at improving the oral and written communication process at the job site with a two fold objective; one of more timely and accurate data and reports, and the other being a means of using the communication process of getting all the job site personnel aligned to the company goals for the project.

## Planning a productive job site layout

One of the more important and often overlooked organization tasks of a construction supervisor is the laying out of the project site. Questions as to where to place trailers, where to store on-site materials, and where equipment should be located at a job site when not in use all necessitate decisions that are part of the site layout task.

All too often a supervisor fails to analyze alternatives when it comes to the layout task. Instead, the supervisor somewhat haphazardly puts the trailer at one location, the material storage area at another, the subcontractor trailers at other locations, and so on. The supervisor may overlook the effect that the assigned locations of these job support components has on productivity, safety, worker satisfaction, and communications.

For a given construction job, there is one and only one optimal layout. If a job is planned on the basis of any other layout, some aspect of the working environment will be less than optimal. For example, if the materials storage location is such that workers have to continually walk long distances to get needed materials, hours of non-productive labor time will result. Similarly, if materials are stored away from the place of their subsequent fabrication, they will need to be unnecessarily double lifted. Storing heaving equipment in the path of workers may increase the chance of injury. Placing a material storage area in an area of no visual control or near the entrance or exit to the job site may promote theft.

Consider the example job site layout illustrated in Figure 11.5. It is not a productive layout in that materials are stored too far from the fabrication area, the trailer is in an ackward place, and equipment is stored in a location that cause congestion and possible accidents.

The job layout illustrated in Figure 11.6 represents a significant improvement relative to the one shown in Figure 11.5. It will improve project productivity, communication, safety, and minimize theft. The point to be made is that expending a few minutes of pre-planning time, an improved job site layout can be prepared.

A structured approach to the job site layout task is needed. Such an approach should recognize "things" and "concerns". The things are various trailers, materials, equipment, storage areas, signs, lavatory locations, luncheon and break area locations, etc. Concerns include distances required to be walked, safety, accounting controls, material or equipment theft, avoidance of adversarial human relations, and overall productivity enhancement.

## Challenging the work process

The role of the contracting supervisor has been more of a policeman than that of an analyst; someone that is always looking for a better way to do things. This is because of the short term nature of the construction process,

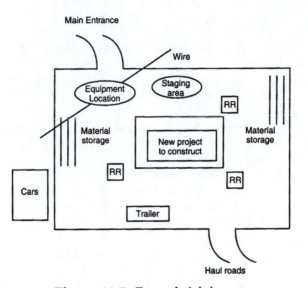

**Figure 11.5.** Example job layout.

and given the fact that it is harder to critique than to watch. His role is one of seeing to it that the plan is carried out; that every worker is doing his part and is worker hard.

The fact remains that it is easier to watch the construction process than it is to critique the work process looking for a better way to do the work. How then can we expect the supervisor to change his ways to become an analyst looking for a more productive work process; one that will yield a lower unit cost of performing the work? A process of simply having the supervisor challenge a work process with the following questions can lead to the identification of a more productive work process.

- Why are we doing it this way?
- Where is the best place to do it?
- When is it the best place to do it?
- Which equipment is best to use?
- Who is best qualified to do the work?
- What are we ultimately trying to accomplish?

The objective of asking these questions is to challenge the current work process and hopefully derive a more productive work process.

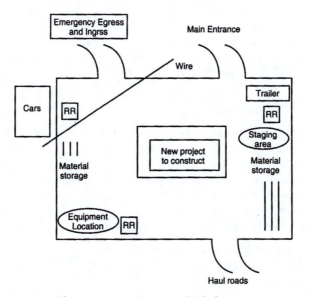

**Figure 11.6.** Improved job layout.

## Developing a scientific work standard versus an accounting based standard

One could propose that in regard to productivity work standards in the contracting industry, inefficiencies become standards. The contractor develops it's work standards and productivity standards for future job estimates by using the accounting process to collect productivity and cost data from in process projects. In an industry that has as much as fifty percent non-productive time in the work process (or fifty percent opportunity to improve), productivity standards developed from past projects clearly do not set out what productivity can be achieved, only what is being achieved.

Many industries develop their productivity standards through the analysis of the work process. Many manufacturing industries make use of industrial engineering models such as time study models, process charts, flow diagrams etc. to study the work process with the intent of determining an achievable work or productivity standard as well as a measurement of the productivity they are currently obtaining.

The fact remains that scientific models such as time study, work sampling, the process chart, and many more have a place in the program for productivity improvement. If the contracting supervisor uses these models with common sense and promotes them as techniques for determining improved productivity standards that can be accomplished by working smarter not harder, they can become a practical and highly beneficial means of setting out better ways to do things. The supervisor should seek to learn and apply various models for analyzing productivity to include time study, work sampling, the process chart, the flow diagram, and several other models that can be found in the literature.

## Planning and scheduling

The supervisor does not always prepare a formal plan and schedule. There is little doubt that the industry has made more use of formalized schedules such as the critical path method (CPM). This use has been enhanced by the availability of many computer software programs that have lowered the time and cost to prepare such schedules.

There are actually three somewhat different types of plans or schedules that can be used by the supervisor to improve on-site productivity.

1. Short interval planning or a one day plan
2. One- to three-week revolving plan and schedule
3. Master schedule (for example CPM)

The short interval schedule form was discussed in the planning and scheduling chapter. Job site productivity can be significantly increased if the supervisor would plan the next day's work one day ahead of time. This could be accomplished in a formal way if he made use of a form at the end of each day that required him to set out each of the following:

1. What type of work is to be accomplished tomorrow.
2. A work quantity goal for each type of work to be performed tomorrow.
3. What types of resources to include labor, tools, material, and equipment will be needed tomorrow to accomplish the set out work goals.

The idea is to ready the resources before they are needed. This process can play a significant role in reducing or eliminating non-productive time related to the waiting time that characterizes many work processes.

An example one- to three-week look ahead scheduling form is illustrated in Figure 11.7. The use of a one to three week revolving schedule that sets out work goals has the same objective as the one day short interval schedule. The difference is that by setting out what work will be done over the next few weeks, the supervisor can start the process of securing tools, material, labor, or equipment that will have a few days or few weeks lead time in procuring. For example, it may take three or four days to obtain certain types of material needed for the work process.

| # | Tasks | Mat'l | Tools | Crew | Ctr | Week 1 | Week 2 | Week 3 |
|---|-------|-------|-------|------|-----|--------|--------|--------|
|   |       |       |       |      |     |        |        |        |
|   |       |       |       |      |     |        |        |        |
|   |       |       |       |      |     |        |        |        |
|   |       |       |       |      |     |        |        |        |
|   |       |       |       |      |     |        |        |        |
|   |       |       |       |      |     |        |        |        |
|   |       |       |       |      |     |        |        |        |
|   |       |       |       |      |     |        |        |        |

**Figure 11.7.** One- to three-week look ahead schedule.

A master schedule can be prepared using CPM computer software. The benefits of CPM were discussed in a prior chapter. In summary, the master schedule provides a road map for the construction project. It helps the user make decisions aimed at minimizing the project time and cost.

## Project control and productivity improvement

One might propose that if you don't know you have a problem, you can't address the problem for improvement. When one looks at the job cost reports that characterize some contracting jobs or firms, one could suggest that they are not timely and may be inaccurate. If manhour productivity for doing work such as manhours required to place concrete forms is going over budget, the time to address the productivity manhour overrun is while the work is being performed, not when the work is complete.

The contracting supervisor needs to implement a cost accounting system that gives a timely report on the percentage of effort expended versus the percentage of labor or equipment effort expended for each and every significant work task that needs to be performed.

Assume that the report indicates that the percentage of labor hours expended are exceeding the percentage of work in place. There could be several reasons for this that are not correctable to include the preparation of an inaccurate estimate, inaccurate field reporting, or unexpected and uncontrollable job site conditions such as adverse weather conditions. However, one of the reasons could be a correctable productivity problem such as having an ineffective crew combination, poor supervision, a lack of an understanding of the work, etc. The point to be made is that by the report drawing attention to the mismatch between work performed and labor effort expended, the supervisor should be able to investigate what may be a correctable productivity problem. The time to address the productivity problem is while it is correctable, not after the work is complete and results in a non-correctable cost overrun.

## Reducing multiple material handling

A major defect in the construction process is the double, triple, or even more handling of material at the job site. Placing a rebar at one location, only to move it to a different job-site storage location, then moving it again to the point of placement, is an example of an inefficient production process. In addition, this multiple handling of materials increases the chances for worker injury and defective material.

Studies indicate that the average construction material is handled three plus times at the job site. The goal should be to handle it once. Attention to a good job-site layout, timely ordering of material, and a plan and strategy for material placement can promote this "handle it once" objective. For one, a small piece of paper can be attached to every piece of material as it is brought to the job site. Anyone moving or handling the material can be required to initial the paper. The goal should be to find a way to minimize handling such that there is a single set of initials on each material.

## Productivity improvement through safety

Studies have proven that the construction industry to include the contracting firm has one of the highest worker accident rates per number of worker hours expended. This is due in part to the difficulty of the work and the conditions in which many projects are constructed.

Regardless of the reasons for the many construction accidents that occur at job sites, the fact remains that they have an adverse affect on construction productivity. In addition to the detrimental effect of the injury for the worker himself, accidents are likely to cause low worker morale, work disruptions related to identifying the cause of the accident, and higher insurance premiums.

More often than not, a productive job is a safe job. A worker is as likely or more likely to get hurt when he is non-productive versus when he is performing productive work. A worker is a state of boredom or in lackadaisical state may find his mind wandering or be careless to the point he puts himself in an accident prone situation. An effective safety program that complies with safety regulations and promotes safety to the workers, is compatible with the firm's productivity improvement program.

## Productivity through attention to quality

Last but not least in importance to a program for productivity improvement is the need to be attentive to performing high quality work. Poor quality construction can negatively impact productivity improvement program by (1) creating an environment in which workers know that a less than desirable quality is accepted and therefore perhaps a less than good work effort is also acceptable, (2) a tendency to have the worker lose pride in his work effort, and (3) the possibility for the need to do redo work that directly increases the required number of worker hours to do the finished work.

By giving attention to making the project safe and attention to the performance of high quality work, the supervisor can serve the objective of making a project look like a firm to the worker rather than just another job. Supervisor efforts to keep a project safe and efforts made to maintain high quality will lead to increased productivity.

# 11.5 Increasing Construction Equipment Productivity

Contractors pay more attention to labor productivity than to equipment productivity. This stems from the fact that contracting firms often view workers as an hourly cost. However, construction equipment — such as cranes, excavators, concrete pumps and pavers — also has an hourly cost.

Various studies indicate that on average, that workers are productive about 50% of an eight-hour workday. By productive, we mean doing necessary and value-added project work such as erecting concrete forms or placing and finishing concrete. During the rest of the eight-hour day, workers are engaged in non-productive activities such as waiting on materials, waiting for instructions from supervisors, redoing unacceptable work or looking for tools. Although this percentage of non-productive labor hours certainly should be a concern to contracting firms, let's compare it to the average percentage of non-productive hours for construction equipment

According to leading equipment manufacturers, contractors can expect to get 800 to 900 productive hours from a typical piece of equipment in a given year. Assuming a 2,080-hour work year (52 weeks, with a 40 hour work week) and 800 hours of productive equipment time, construction equipment is non-productive 62% of the time ((2,080 − 800)/2,080), which is significantly higher than the 50% non-productive time for workers.

The contractor should take steps to improve both labor *and* equipment productivity. Following are ways you can increase productivity by improving equipment management.

## Keep Track of Productive, Standby, and Non-Productive Time

Like a worker, a piece of equipment at a jobsite is either working or not working. To improve equipment productivity, the contractor should keep daily records of how equipment time is being used. The firm should keep the percentage of time each major piece of equipment is in one of the following states.

- Productive (performing necessary work)
- Standby (able to do work, but no work is available)
- Non-productive (not capable of doing the work, broken, or doing unnecessary or low value tasks)

By comparing these percentages the contractor can determine how effective supervisors are using equipment at different jobsites.

The data in Figure 11.8, shows a concrete pump's work percentage for two different jobs. By comparing the data, a contractor can see the standby and non-productive times for the pump are greater on Job 1 than on Job 2. By investigating the causes of these differences, the contractor can take steps to increase productivity.

| Work state of concrete pump | Job 101 | Job 102 |
|---|---|---|
| Productive | 42% | 62% |
| Standby | 30% | 16% |
| Non-productivity | 28% | 22% |

**Figure 11.8.** Example equipment states.

For equipment-intensive jobs, the ratio of productive to non-productive equipment time can be one of the best measures of supervisor performance. The contractor may consider equipment productivity results as a determining factor for supervisor bonuses and profit sharing.

## Develop Accounting Process for Owned Equipment to Enable Jobs to Get Charged for Use of Equipment

All too often, when using their owned equipment for projects, the contractor's accounting process is such that the job does not get charged for the use of the equipment; instead the job only get charged for any equipment related expenses that occur while the equipment is at the job. The end result is that some jobs get overcharged for equipment expenditures and other jobs get undercharged. In addition, the contractor does not establish the true cost of owning a piece of equipment.

One of the primary principles of proper cost accounting is that an individual should be accountable for costs or events that they control. A supervisor should not be responsible for costs or events they do not control. The construction supervisor is more in control of the use of equipment at their job site than they are of the actual expenditures of the equipment.

For major pieces of equipment, the contractor should consider implementing an equipment estimating and control system that captures the hourly cost for owning a specific piece of equipment and also charges projects for the specific use of the equipment. The hourly cost of owning a piece of equipment is shown in Figure 11.9. Depreciation is the means

$$
\begin{aligned}
\text{Hourly cost} = \ &\text{Depreciation} \\
&+ \text{Maintenance} \\
&+ \text{Operating} \\
&+ \text{Repairs} \\
&+ \text{Finance costs} \\
&+ \text{Insurance costs} \\
&+ \text{Property tax} \\
&+ \text{Replacement cost}
\end{aligned}
$$

**Figure 11.9.** Hourly cost.

by which the initial purchase cost is recovered and charged to projects. Because equipment normally wears out as a function of use (rather than time), it is better to use a depreciation method that is a function of use. In determining the hourly depreciation rate, the contractor must consider the initial cost of the equipment, the expected salvage value, and the expected hours of usage.

An example of a maintenance cost would be routine equipment maintenance such as an oil change. Repair costs would include undercarriage repair, and costs to repair a damaged motor. Operating costs would include gasoline and perhaps the operator cost (assuming the operator is viewed as being attached to the equipment).

The finance cost would be interest expended related to borrowing money to purchase a $100,000 piece of equipment or the loss of opportunity of interest income associated with using one's own funds to purchase the $100,000 piece of equipment. In either case, the cost of owning the equipment should include the interest cost on value of the equipment.

Similar to the above noted finance cost is the insurance cost that the contractor likely carries on a specific piece of equipment. Assuming that the annual insurance cost for our example $100,000 piece of equipment is $2,000 and that it is estimated that it will be used 800 hours in a year, the insurance cost would be $2.50 per hour.

Some states continue to have property tax assessments for assets such as equipment. If this is the case, this cost component is part of the cost of owning equipment.

The replacement cost component shown in Figure 11.9 relates to the fact that the depreciation by itself does not set aside enough to replace a piece of equipment when it wears out. Equipment manufacturers may be increasing the cost of new equipment by 2–5 percent in a given year. If the contractor does not take this into consideration when charging a project owner for his equipment, one might refer to the contractor as going equipment poor. The contractor should include a cost component in the hourly equipment rate that sets aside a reserve to replace the equipment at an increased purchase cost.

Given the knowledge of the above equipment ownership costs, we will now describe an accounting process that promotes equipment productivity improvement.

1. Start with a manufacturer prescribed rate.

   Because the contractor starts with a lack of historical data regarding the cost of ownership (for example, no maintenance cost history) the process starts with the use of an average ownership rate as prescribed by the equipment manufacturer or as suggested in industry cost books. For our example $100,000 piece of equipment, let us assume the rate is $50.00 per hour.

2. Determine estimate for specific project.

   Assume our example equipment is to be used for a project being estimated. The estimator determines that it will be needed for 200 hours of use on a new project, Job 101. The estimate should include the equipment cost of $50 per hour times 200 hours or $10,000.

3. Job 101 gets charged for use of equipment.

   As the job progresses, it should be charged for the use of the equipment. For example, at the end of the first month of the job, the equipment may have been used 30 hours. The job ledger for Job 101 should reflect an expense of $1,500 (30 hours times $50 per hour). The point is that the job and the supervisor gets charged for the use of the equipment.

4. Equipment expenditures get charged to an equipment account for the equipment.

   As actual expenditures are incurred for the equipment in our example, they are not charged to specific jobs, they are kept track of on an individual piece of equipment basis. This is done by debiting a general ledger account set up for the equipment. Some of these expense debits would be periodic like interest, and insurance expenditures. Other expenses would be debited or charged as they were incurred such as maintenance, and operating cost. However, these and all actual expenditures are charged to the ledger account for the equipment, not to the specific project.

   The process described in Steps 3 and 4 would continue throughout the year. Jobs would be charged for the use of the equipment and all equipment expenditures would be captured by the equipment account.

5. Determine actual equipment cost from company records.

As noted in Step 1 the process starts with the assumption of an equipment prescribed average hourly rate or ownership; in our example $50.00 per hour. However, this is not the actual cost of ownership to a specific contractor. The actual rate is determined by analyzing the captured actual equipment expenditures and the actual hours of use. The hours of use are determined as a by product of Step 3 noted above. When a job such as our Job 101 gets charged for the use of the equipment (by means of debiting the job cost), an offsetting credit entry is made to the machine ledger account. For all practical purposes, the process works like the accounting system is renting the equipment to specific jobs; the credits are made to the equipment account when the equipment is used at jobs., and debits are made to the equipment account when actual equipment expenditures are incurred. An overview of this process is illustrated in Figure 11.10.

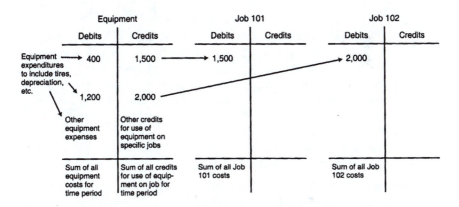

**Figure 11.10.** Equipment estimating system.

At the end of a specific time period, say a year, the equipment account would be analyzed. If the debit entries (which represent the actual expenditures) exceed the credit entries (which represent the charges made to jobs for the use of the equipment), this implies that the actual ownership cost per hour was in excess of $50.00 per hour. It might also mean that equipment was not used enough at projects during the year relative to what was anticipated.

By having also captured the actual hours of usage during the time period for which equipment expenses were tabulated, the contractor can proceed to calculate his own cost of ownership. The point to be made is that the above process holds individuals responsible for the use of equipment. Accountability forces responsibility and ideally, productivity increases. The contractor supervisor controls the use of equipment and should be held accountable for the use of equipment. The equipment accounting system described does this.

The difficulty of the above described process is the increased paperwork that is associated with keeping track of the actual use of equipment on jobs, and the paperwork associated with tracing equipment expenditures back to the specific piece of equipment. However, with good paperwork procedures, this process can be done efficiently.

To reduce the paperwork associated with the above process, some contracting firms alleviate some of the paperwork by creating "pools" of equipment. For example, all cranes of a similar type may be handled by an aggregated crane account. In effect, the hourly rate jobs are charged at a "pooled rate".

## Consider the Hourly Cost of Owning or Renting Equipment

To effectively manage labor and equipment, you must know what they cost. Most supervisors know workers' wage rates, but they may not know the hourly rates for major equipment such as a concrete pump or a crane.

Equipment is expensive. Example hourly rates for equipment include a concrete paving at $180 per hour or a pump at $90 per hour. To encourage construction personnel to view equipment as money, not just a machine, try posting or painting the hourly cost of renting or owning equipment right on the equipment itself, visible to all. Following are other ways you can make employees more aware of equipment rates:

- Show employees the hourly equipment costs used in the estimate and share information about actual equipment costs vs. budgeted amounts.
- Make equipment part of the hourly cost of a work crew, so employees can see the impact on the bottom line.
- Post hourly costs of expensive equipment near the jobsite trailer or entrance to stress to workers that idle equipment is wasted money.

The bottom line is to get supervisors to view equipment as a cost, not just a piece of metal. Idle equipment eats at the bottom line profits, just as does non-productive labor time.

## Measuring and Managing Equipment Risk

Just as it is important to view a piece of equipment as having an hourly rate, it is also important to view a piece of equipment as an element of project risk. Risk can be thought of as possible variation from expected results. Consider the two construction methods illustrated in Figure 11.11. Assume that a construction supervisor has collected past project data regarding two alternative equipment ways of placing concrete, one with a pump and one with a crane bucket. The points shown in Figure 11.11 under the two bell shaped curves represent the amount of concrete placed per dollar of cost which includes the cost of the two different pieces of equipment. Each point represents the performance on a past project.

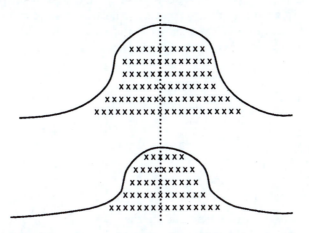

**Figure 11.11.** Two construction methods and risk.

Using bell shaped curves representative of what is referred to as a normal distribution in probability theory, the past samples are grouped by the curves shown. While the past project data implies that both methods (each using a different piece of equipment) have the same expected average unit cost (or productivity), the past project data implies that the samples shown as Method 1 have varied more from the average than those of Method 2, Method 1 has more productivity risk.

In selecting and monitoring equipment, the supervisor must pay attention to productivity risk as well as average. In the example, if the equipment utilized is represented by the data as shown in Method 1, the supervisor needs to closely monitor the equipment and work process; it is subject to considerable risk. On the other hand, given the wide variation of productivity achieved, as represented by the large bell shaped curve in Method 1, close supervision may in fact lead to a higher productivity than that achieved with the equipment used for Method 2. If risk is managed, high productivity can be achieved. However, if risk is not managed, the results can lead to disastrous productivity.

The point to be made is that in selecting and monitoring equipment performance and productivity, the contractor needs to manage risk as well as managing things. By focusing on the hourly cost and the hourly risk of equipment, productivity can be managed and improved.

## Monitoring Variable Versus Fixed Equipment Costs

The lack of an understanding of equipment costs can relate to a contractor's inability to hold individuals accountable for costs and may lead to the contractor's failure to recover the costs for a change order or dispute. To illustrate the importance of understanding the true cost of construction equipment, let us consider the issue of a work change or dispute which includes the incurrence of equipment costs. The cost components of owing equipment are:

- Depreciation
- Maintenance
- Repair
- Operating
- Finance cost
- Insurance cost
- Property tax (if applicable)
- Replacement cost

Two separate situations may initiate the issue of fixed versus variable costs. For one, the contractor may have had to do added work; i.e., change order work. The project owner may take the position that as long as the project took the same duration as what was planned, that the contractor is not entitled to added costs for his owned equipment. The owner's position is predicated on the assumption that the owned equipment cost is a sunk cost; i.e., the equipment was there anyway.

The second situation relates to a delay in a project that required the contractor to have the equipment at the project longer than anticipated. The owner may take the position that unless the contractor can show that there was another project waiting the availability of the equipment, that the equipment is a sunk cost and therefore the contractor is not entitled to added compensation.

Equipment costs can be viewed as either fixed (function of time only) or variable (function of equipment use and activity). The following costs are primarily a function of use:

- Depreciation (Note: Equipment wears out more as a function of time.)
- Maintenance
- Repair
- Operating

On the other hand, the following equipment expenditures are primarily a function of time:

- Finance cost
- Interest cost
- Property tax (if applicable)

• Replacement cost

The contractor incurs these costs independent of whether or not a specific piece of equipment is doing actual work; they are a function of time.

If the contractor and supervisor knows each of the above cost components by having kept records, the contractor can justify requesting cost components that are a function of use when the contractor has an added work but not added time issue. Similarly, when the contractor is delayed with his equipment, even though the contractor did not do any added work, he can make a strong case for the cost components that are a function of time.

In addition to the issue of equitably resolving the issues of change orders and claims, the knowledge of variable and fixed expense components enables the contractor to make optimal buy versus rent equipment decisions, to measure equipment productivity, and to hold appropriate personnel responsible for the cost and time use of equipment. One of the keys to productivity improvement is accountability.

## Scheduling of Equipment

The productive use of construction equipment starts with the scheduling of the equipment. Construction equipment should be scheduled and monitored for a project the same way labor is. Using a manual or computerized scheduling program, it is possible to equipment load the schedule in the same manner the schedule is manloaded. Planning and scheduling was presented in another chapter.

The objective is to preplan the use of equipment prior to the project. By analyzing an equipment loaded schedule and by using schedule float, it is often possible to sequence project work to make improved use of equipment. For example, if a project is scheduled such that a crane or concrete pump is only needed two out of every five days in a week, at best, the contractor can expect forty percent productivity with the equipment (2 days/5 days). If the schedule can be resequenced and monitored such that the equipment will be needed three days in a given week, this will raise the potential for productivity to sixty percent (3 days/5 days). The end result is that through effective scheduling, a fifty percent increase in equipment productivity can be attained (from forty to sixty percent potential productive time; a fifty percent increase). Effective scheduling of labor and equipment provides a means of attaining productivity increases by working smarter, not harder.

## Managing Maintenance and Repair

Given the relatively large expenditure that one makes for construction equipment (often in excess of $100,000 for a single piece of equipment), it is no wonder that one may overlook the importance of maintenance and repair costs. However, if one analyzes the true cost of owning equipment, one would find that the average hourly cost of maintenance and repair cost may exceed that of the purchase cost broken down into a hourly depreciation cost. For example, one of the leading manufacturers of equipment breaks down a typical hourly ownership of a $100,000 piece of equipment as follows:

| | |
|---|---|
| Depreciation (purchase cost-salvage component) | $8.00 |
| Maintenance cost | $7.00 |
| Repair cost | $8.00 |
| Other cost components (finance, etc) | $23.00 |
| Total | $46.00 |

While maintenance and repair costs are expensive components of construction equipment ownership, of more importance is the added costs that the contractor can incur because of a lack of proper maintenance and repair of equipment. Some of the direct and indirect costs that occur owing to a lack of improper equipment maintenance and repair are the following:

- Idle project resources (e.g., labor) waiting on broken down equipment
- Costs associated with less than optimal equipment performance
- Project delay costs associated with equipment down time
- Costs associated with a construction accident that results because of equipment breakdown
- Major repair costs associated with failure to maintain or expend funds for minor repairs

Non-productive equipment time and the above types of costs can be minimized through the implementation of a cost effective equipment preventive maintenance (PM) program. This entails making maintenance and repairs expenditures to equipment on a regular prearranged schedule, prior to equipment failure. When one considers the cost of non-productive labor and equipment costs associated with an equipment breakdown, these preventive maintenance expenditures are relatively minimal

The high costs associated with the failure to properly maintain equipment are not only large but are also uncertain and subject to considerable risk. Unlike ownership costs (depreciation) that are known and predictable once the ownership and salvage value can be determined, the sporadic and uncertain nature of maintenance and operating cost make the management of the maintenance function even more critical.

The keeping track of hours of usage of equipment and repairs is an important component of the preventive maintenance program. Such records also enable the construction firm to make improved decisions regarding the trading in of a piece of equipment, the time to retire a piece of equipment, and the overall scheduling of equipment to projects.

## Knowing Where Your Equipment Is

Given the hectic nature of the construction process and the contracting firm itself, it is not unusual to witness considerable loss of project productivity, time, and cost while project resources are waiting on a piece of equipment to arrive on the job site. The reason often relates to the fact that the contracting firm, let alone the job supervisor, may not know the whereabouts of specific pieces of equipment. When one calculates the cost of idle labor and equipment costs that may incur owing to waiting or looking for a needed piece of equipment, the hourly cost may be in the hundreds if not thousands of dollars.

A simple solution would appear to be a perpetual inventory system for equipment; a system that tracks where a piece of equipment has been, where it is now, and where it is going next. Such a system may be in the form of a simple blackboard that is updated on a daily or timely basis or may entail the use of a computer software program that tracks each and every piece of equipment the firm owns. The point to be made is that if the contractor does not have a system in place to know where each and every piece of equipment they own or rent is, they are likely to incur significant project productivity losses and also increase the possibility of equipment theft.

## Selecting and Analyzing the Right Equipment for the Job

One of the major causes or factors that leads to low construction industry productivity relates to contractors not using the right equipment for the job. The problem is that the contractor seldom can afford the latest and greatest piece of equipment. Owing to his financial position, the contractor estimating and performing a project, often has to use the equipment they have to do the job, even if it is not the best equipment for the job in regard to productivity or cost effectiveness.

While the contractor finances curtail the use of purchasing the best equipment fleet as new jobs evolve, it is still in the best interest of the contractor to do a cost analysis of the best equipment for each and every project. Sometimes such analysis will lead to renting a more productive piece of equipment for a task even if the contractor owns a piece of equipment that will do the task (but at a higher hourly or unit cost). This is why it is critical that contractors regularly attend trade shows. The very knowledge of new equipment and technology that can increase productivity puts the contractor in a position to evaluate the best equipment for a project. Using outdated, inappropriate, or non-effective equipment to do a work task may lead to low productivity, high costs, time delays, and increase the potential for an accident. All these factors should be considered when selecting the right equipment for the job.

### Productivity Improvement Through Safety

A productive job can also be a safe job and vice versa. In fact one can argue that one cannot have one without the other. Project data often indicates that accidents occur when a worker (or a piece of equipment) is a non-productive work state. There is evidence that more accidents occur when labor or equipment is operating in less than a productive manner than if the worker or machine is operating productively.

Similarly, one cannot have accidents and expect productivity to be high. Job site accidents, be they caused by a worker or a piece of construction equipment, lead to non-productive project time, disarray, and low morale; not to mention the added cost associated with worker injury or equipment and project repairs necessitated by the accident.

Time and cost expended to keep construction equipment located in a safe working area, properly secured, properly positioned, and properly operated, are more than offset by the saving of the high costs associated with accidents. Construction equipment productivity and safety go hand in hand; one cannot achieve one without the other.

## 11.6   Summary

The key to profitability and the minimization of project risk is the increasing of job site productivity. Often as little as a five percent increase in labor or equipment productivity has the result of doubling the profitability of a job and the contractor. In order to increase productivity, the contractor must be proactive instead of only being reactive to job site problems. This includes developing specific practices for labor and equipment productivity improvement.

# Chapter 12

# Job Control:
# Time and Cost

## 12.1  Introduction

Controlling project time and cost are key to the contractor's profitability
and the ability to complete a project on time and budget. While the project
estimate and the project schedule set out the potential for profits and an
on-time project schedule, it is the control firm that enables the contractor
to achieve the profit and schedule.

A control system has five key elements:

1. In a *timely* manner,
2. it compares *actual*,
3. to *plan*,
4. with the objective of *detecting a problem*,
5. and a *follow-up action* is taken to attempt to correct the problem.

If any one of these five elements are absent or are done improperly, the
control process fails.

## 12.2  Job Cost Reporting Process

The objective of the job cost report is to enable the contractor to "see"
the status of an in progress project from the company office. The company
owner and his key company personnel to include project managers cannot
be at each and every job all the time. It is the timely job cost report that
provides the firm the opportunity to see the project from the office.

The largest risk factor, and potentially the largest profit center, for
the contractor is the on-site self-performed labor cost. Two key job site
reports support the preparation of the job cost report; the daily report

that summarizes quantities in place, and the time card that summarizes craft hours worked. This process is shown in Figure 12.1.

| Items taken-off from drawings | Quantities taken-off by estimator | Budgeted hours based on historical data | Quantities to date from daily reports at job | Craft hours to date from daily timecards | Calculated from data in previous 4 columns |
|---|---|---|---|---|---|
| Work Item | Budgeted Quantity | Budgeted Hours | Actual Quantity | Actual Hours | Projected Over or Under Hours |
| | | | | | |
| | | | | | |

**Figure 12.1.** Job cost report.

In order to be effective, the budgeted quantities and craft hours have to be accurate. Equally important, the field reporting system must be timely and accurate. The source documents for the quantities of work put in place to date and the craft hours expended to data are the daily reports and the time cards filled out at the job site. We will return to the need for accurate and timely data in a later section of this chapter.

Contractors use various formats of the job cost report. One format is shown in Figure 12.2. This format can hardly be viewed as being effective. It merely measures the labor hours expended to date against the budget. This enables the contractor to detect a problem when the labor hours exceed the budgeted hours; much too late to be able to make corrective actions. For example, the report shown in Figure 12.2 would enable the contractor to know they have a productivity (and cost overrun) problem with forming slabs when in fact the actual craft hours exceeded the budgeted 500 crafts hours.

| Work item | Budgeted quantity | Budgeted craft hours | Actual craft hours |
|---|---|---|---|
| Form slabs | 500 | 500 | 350 |
| Place rebar | 200 | 300 | 150 |
| Place concrete | 100 | 200 | 80 |

**Figure 12.2.** Comparing actual hours to date versus budgeted hours.

A more effective control report that monitors percentage of effort or labor hours expended versus the percentage of work put in place is illustrated in Figure 12.3. In Figure 12.3, 500 units of work (forming is usually measured in square feet of contact area) and 500 craft hours are budgeted for forming slabs. At the time of the job report shown, 250 units of work are in place. This represents fifty percent of the budgeted quantity. However, the firm has already expended 350 craft hours to do the work; this represents seventy percent of the budgeted craft hours.

Looked at another way, it has taken 350 craft hours to place fifty percent of the quantity of work. If this productivity (man-hours per unit) trend continues, it follows that it will take a total of 700 craft hours to complete all the work. This is 200 more man-hours than budgeted. This 200 overrun variance is projected in the last column of Figure 12.3.

A similar analysis of comparing the percent of quantity in place and hours expected to date indicates that the placing of rebar work item is right on budget. The placing of concrete work item is projected to come under budget by 40 craft hours.

Using the control system report shown in Figure 12.3, the contractor can detect a potential problem as soon as the percentage of labor hours expended is greater than the percentage of work put in place. This potential problem may occur as soon as the work is barely started. Early detection of a productivity problem is critical to the contractor's ability to correct a problem.

| Work item | Budgeted quantity | Budgeted hours | Actual quantity to date | Quantity in place (%) | Actual hours to date | Hours to date (%) | Forecast variance at completion |
|---|---|---|---|---|---|---|---|
| Form slabs | 500 | 500 | 250 | 50 | 350 | 70 | +200 hrs |
| Place rebar | 200 | 300 | 100 | 50 | 150 | 50 | |
| Place concrete | 100 | 200 | 50 | 50 | 80 | 40 | −40 hrs |

**Figure 12.3.** Job control by comparing percent of hours to percent of work in place.

This comparison of the percentage of work put in place versus the percentage of effort expended to date can be made by comparing hours, labor costs, or unit costs. A report illustrating labor dollars instead of labor hours is illustrated in Figure 12.4. An hourly rate of $20 per hour is assumed in Figure 12.4. The variances in the last column are calculated the same way as they were in Figure 12.3. For example, it has taken $7,000 of labor cost to do fifty percent of the work item "Form Slabs". If the productivity stays the same, it will take $14,000 to complete the work; an overrun of $4,000.

| Work item | Budgeted quantity | Budgeted labor dollars | Actual quantity to date | Quantity in place (%) | Actual labor dollars | Dollars to date (%) | Forecast variance at completion |
|---|---|---|---|---|---|---|---|
| Form slabs | 500 | $10,000 | 250 | 50 | $7,000 | 70 | +$4,000 |
| Place rebar | 200 | $6,000 | 100 | 50 | $3,000 | 50 | |
| Place concrete | 100 | $4,000 | 50 | 50 | $1,600 | 40 | −$800 |

**Figure 12.4.** Job control by comparing percent of dollars expended to percent of work in place.

A similar report using unit costs is illustrated in Figure 12.5. The $20 unit cost for the work item "Form Slabs" is determined in the estimating process by dividing the $10,000 budget (i.e., 500 craft hours at $20 per hour), by the budgeted 500 units of work to do. The actual unit cost of $28 for the "Form Slabs" work item at the date of the report is determined by dividing the \$7,000 cost to date by the 250 units of work put in place to date. Using this report, any time the actual unit cost gets larger than the budgeted unit cost, a "flag" goes up that an apparent problem exists.

| Work item | Budgeted quantity | Budgeted unit cost | Actual quantity to date | Actual unit cost to date | Forecast variance at completion |
|---|---|---|---|---|---|
| Form slabs | 500 | \$20.00 | 250 | \$28.00 | +\$4,000 |
| Place rebar | 200 | \$30.00 | 100 | \$30.00 | |
| Place concrete | 100 | \$40.00 | 50 | \$32.00 | −\$800 |

**Figure 12.5.** Job control by comparing unit costs.

The job cost report formats illustrated in Figures 12.3, 12.4, and 12.5 have the ability to satisfy the first four elements of an effective control system. These control elements are:

1. A *timely,*
2. comparison of *actual,*
3. vs. *plan* (or budget),
4. to detect, a *potential problem.*

When the percentage of effort for a work activity is greater than the percentage of work completed, there are at least six explanations or reasons. If the control system is to be effective, it is critical that the contractor investigate the reasons for the overrun. Each of the reasons listed below should result in a follow up action.

1. Productivity problem.

   It may be that there is one of many productivity problems occurring; for example, inadequate supervision, understaffing of the work, poor worker attitudes, etc. If this is the reason, the project control report identifies or "flags" this and the contractor should attempt to immediately correct the productivity problem. Some problems like inadequate or improper supervision can be addressed. Other problems, like inclement weather may be more difficult to address.

2. Estimating problem.

   It may be that there is no on-site problem; the problem may be that there was too optimistic of an estimate of labor productivity. If this is the case, the construction firm should recognize this fact and incorporate this information into future estimates.

3. Inadequate record keeping.

   The fact that the report indicates more hours expended for the Form Slabs than should have occurred to date may be the result of intentional or unintentional inaccurate job site record keeping. For example, it may be that the foreman has filled out daily timecards inaccurately by charging craft hours to the wrong work code. If this is the case, the record keeping process should be improved. Record keeping is addressed in following sections of this chapter.

4. Improper method of handling change orders.

   The contractor is often required to do additional work via a change order process. For example, additional concrete work may have been assigned to the firm. In doing this work, the contractor would update their actual labor hours and work quantities performed. However, they may not have updated the estimate and therefore the percentages illustrated for the labor hours expended and quantity of work placed are in error. If this is the reason for the apparent lack of matching of the percentages, the change order process should be corrected; the budget and the actual quantities and hours must be updated for change orders.

5. Improper or bad list of work item codes.

   The job control report should be set up that if the contractor is fifty percent done with a work item and everything is going as planned, then fifty percent of the labor hours should have been expended. It may be that the reason for the apparent problem with the Form Slab work item is that the firm has a bad list of work items. Maybe the work codes are defined too broadly. If this is the case, the firm should redefine the work codes in the cost system.

6. Changed work conditions or a claim.

   It may be that the construction firm is being required to do the concrete wall work under changed or unexpected work conditions (for example; the work is being impacted by an obstruction). The contractor may be entitled to extra payment for the changed condition or may initiate a claim. If this is the case, this should be documented such that the claim can be negotiated. The job control report can support this with documentation.

The job cost control report is only effective if overruns are investigated and attempts are made to correct overruns. One way the contractor can force the "reaction" to the job cost report would be to require the on-site supervisor to formally respond (with a short written report) as soon as the variance is projected.

## 12.3  Focusing on Under Budget Work Items and Variation

It is also important for the contractor to investigate underruns as well as overruns. It is fairly common for a contractor to place considerable emphasis on investigating problems, identifying the blame, and reprimanding the guilty individual. However, it is also important to spend time identifying reasons while various work items on the job control report are under budget.

For example, for the Place Concrete work item in Figure 12.3, the item is projected to come under budget by 40 man-hours. Just like there are reasons why a work item is over budget, there has to be a reason why the work item is under budget. By spending time with the supervisor that beats the budget for various work items, the cause can be identified and attempts made to duplicate the success. The contractor should attempt to eliminate reasons for problems and duplicate successes. The job control report can be useful in meeting both of the objectives.

In addition to aiding control, the job control report can also aid the contractor by measuring the risk of various work it performs. Risk can be defined as variation from the average. A column can be added to the job control report shown in Figure 12.3 that reports the productivity achieved this week relative to either that which was achieved last week or the average achieved to date. The reader of the report will soon be able to sense the productivity risk for individual work items. The productivity variation from one week to another will likely be much higher for work such as forming concrete than for work such as placing the rebar or concrete. It is important for the contractor to be aware of productivity and/or cost risk such that they can assign the appropriate degree of supervision to the work task.

## 12.4    Extending the Job Cost Report to Monitor Project Time

The job control report can be extended to monitor project time as well as cost. This can be done by budgeting and measuring actual crew size per work item or activity as well as quantity of work and labor hours. Consider the data shown in Figure 12.3. By adding the budgeted crew sizes and the actual crew sizes as shown in Figure 12.6, the contractor has a means of monitoring project time and cost.

The format of this job control form enables the forecasting of the variation of activity hours (or cost) and duration. For example, for the forming activity shown in Figure 12.6, there is a forecasted overrun in hours or cost in that based on the work being 50 percent complete, it will take 700 hours to complete all the work; therefore a 200 hour overrun relative to the 500 hour budget.

On the other hand, owing to the larger than planned crew for the forming activity shown in Figure 12.6, there is a forecasted variance of −3.75 days (the work is projected to take 3.75 days less than budgeted). The above analysis assumes an eight hour work day. Given a crew size of ten workers (instead of the budgeted 5), the workers will expend 80 hours per day. Dividing this into the forecasted total 700 man-hours of work, a projected duration of 8.75 days results. The end result is that while a man-hour or cost overrun is projected, the analysis indicates the work will actually take less time than budgeted.

It is important for the contractor to monitor project time and cost for each work item. Just because a specific work item is on or under budget in regard to labor hours or labor cost, does not necessarily mean it will be finished within the alloted time budget as indicated on the project schedule. If the work item is behind schedule, the overall project will be impacted leading to an added cost (for example, the project may be delayed and the contractor will incur a liquidated damage cost). As a work item or activity is in progress, one of four situations is possible:

- The activity can be on or under the cost budget, and be on or under the time budget
- The activity can be over the cost budget, and over the time budget
- The activity can be over the cost budget, and on or under the time budget
- The activity can be on or under the cost budget, and over the time budget

In the later three situations, either a time or cost problem exists. The contractor should react to the problem and implement immediate procedures to reduce or eliminate the problem.

The above extension of the job cost report is only possible if the job cost "work items" are defined to be the same as the project schedule activities and vice versa. This integrating of the job cost system and the scheduling system is discussed in the following section.

# 12.5  Integrating Estimating, Scheduling, and the Control Function

To gain efficiency and accuracy in performing the estimating, scheduling, and control functions, an integrated approach should be taken; i.e., each of the functions should be performed in conjunction with one another by using a common set of work packages. Whatever is a work item for estimating, should be an activity on the project schedule, and should be a cost object in the control system.

This integrated system will now be illustrated via an example of focusing on the work package of "placing suspended floor forms".

## Historical Data Base

We will assume that our example contracting firm has performed previous projects on which they have collected data regarding the placement of slab forms. In particular, we will assume the historical data shown below:

| Project performed | Quantity of suspended floor forms | Crew size utilized | Man-hours expended | Man-hours/ square foot | Cumulative mh/sfca |
|---|---|---|---|---|---|
| School 101 | 10,000 sfca | 20 | 1,000 | 0.10 | 0.10 |
| Hospital 102 | 12,000 sfca | 12 | 1,050 | 0.0875 | 0.0932 |
| School 103 | 20,000 sfca | 15 | 1,600 | 0.08 | 0.0869 |
| Office 104 | 9,000 sfca | 13 | 800 | 0.0888 | 0.869 |
| School 105 | 20,500 sfca | 25 | 2,700 | 0.132 | 0.100 |
|  | 71,500 sfca |  | 7,150 |  | 0.100 |

## Determine the Quantity of Work for the New Project

Let us assume our contracting firm is starting a new project, School 106. The estimator would systematically review the drawings and specifications and calculate the amount of work to be performed for the various defined work items. The calculations and calculated quantities of work are placed on an estimating form or sheet.

While a project would require that several different work items be performed for a project, we will focus on one, Place Suspended Floor Forms. Let us assume that after doing the take-off process, the estimator has determined the following quantity of work.

Place suspended floor forms = 20,000 square feet of contact area (sfca)

## Review Past Data, Select Method, and Determine Man-Hours Required

As a basis of planning the method and determining the cost and duration, the contractor should review past historical data, determine the crew size to be used, and determine the required man-hours that will be needed.

The review of the past data shown indicates that the man-hours required per square foot of contact area of forms placed has varied between 0.08 man-hours per sfca and 0.13 man-hours per sfca. In order to help in determining an estimated man-hours per sfca for the work to be done for School 106, the contractor would decide on how many workers they are going to assign to the work. Let us assume that based on a study of the historical data and the work to be done, the contractor decides on the following:

Crew size = 20 workers

Man-hours/sfca = 0.10 man-hours/sfca

Given the estimate of 0.10 man-hours/sfca, it follows that the total man-hours that will be required for School 106 is calculated as follows:

Man-hours required = (20,000 sfca) × (0.10 man-hours/sfca) = 2,000 man-hours

## Determine Planned or Estimated Cost

Assuming one knows the hourly wage rate for a worker, the labor cost can easily be determined by multiplying the required man-hours times the wage rate. Assume the wage rate is $21.00 per hour.

Estimate labor cost = (2,000 man-hours) × ($21.00/hr) = $42,000.00

## Determining Planned Duration and Project Planning and Scheduling

If the quantity of work, the estimated productivity, and the crew size are known (or estimated), the activity duration can be determined as follows:

Duration = (20,000 sfca) × (0.10 man-hour/sfca)/(20 hours/man-hour)

= 100 hours

Duration = (100 hours) / (8 hours/day)

= 12.5 days (say 13 days)

The duration for each and every activity would be determined in a similar way. The point to be made is that the schedule durations are a by product of the estimating process; scheduling and estimating are done in conjunction with one another.

## Project Control

As the project progresses, productivity is measured against the project plan in an attempt to control the project. This can be done by comparing the actual man-hours per unit of work performed versus the planned or estimated man-hours per unit of work.

To illustrate this process, let us assume that after several days, job site records indicate the following:

Quantity of work in place   =   5,000 sfca
Man-hours expended          =   1,000 man-hours

The contractor supervisor can monitor the work and productivity by either comparing the percentage of work completed to the percentage of man-hours expended, or by comparing the man-hours per unit of work performed versus the budgeted man-hours per unit of work performed.

$$\text{Percentage of work put in place} = \frac{\text{Quantity of work put in place}}{\text{Budgeted quantity}}$$

$$\text{Percentage of work put in place} = \frac{5,000 \text{ sfca}}{20,000 \text{ sfca}} = 0.25 \quad \text{or} \quad 25\%$$

$$\text{Percentage of effort expended} = \frac{\text{Expended man-hours to date}}{\text{Budgeted man-hours}}$$

$$\text{Percentage of effort expended} = \frac{1,000 \text{ man-hours}}{2,000 \text{ man-hours}} = 0.50 \quad \text{or} \quad 50\%$$

The fact that it has taken 50 percent of the budgeted man-hour budget to do 25 percent of the work will alert the contractor to a productivity problem. Attention should be drawn to the apparent productivity problem. Through analysis of the problem, it may be possible to correct the problem.

## Update Past Project Data

The control process described above would be on-going. Many firms generate a control report weekly or monthly to monitor the productivity. When the project is completed, the past project productivity data would be updated to reflect the results of the most recent project.

Let us assume that School 106 is completed. Based on job site records, the final quantities and man-hours expended are as follows:

Quantity of work performed = 20,000 sfca

Man-hours expended = 3,000 man-hours

The past project productivity data would be updated as illustrated in Figure 12.7.

Given another project to plan and estimate, the process described starts again. Past project data becomes the data base for planning and estimating new projects. Estimating, planning and scheduling, and control become part of an integrated project management system.

## 12.6    Input or Performance Management

This chapter has emphasized the control of project time and cost by comparing actual performance against the budgeted performance. In particular, as shown in Figure 12.3 the percentage of effort (expressed in craft labor hours) is compared to the percentage of work put in place. Such a system is much more effective than merely comparing labor hours against the budgeted labor hours. However, both systems can be viewed as being "results" oriented. Attention is drawn to a problem after the problem occurs.

An alternative to the typical job control system is what the author refers to as input or performance management or control. In this system of control, the contractor shifts from results to setting out and monitoring input. For each work item or work task, the contractor would set out the crew size required to do the work. The control focus would then monitor daily the actual crew size.

For example, assume for the work item "Form Slab" illustrated in Figure 12.6, 500 craft hours were budgeted. Assume the work was do be performed in 12.5 days, and that a worker is scheduled to work eight hours a day. The craft hours required each day for the 12.5 days is calculated as follows:

$$\text{Craft hours required per day} = 500 \text{ craft-hours}/12.5 \text{ days}$$
$$= 40 \text{ craft hours per day}$$

The number of workers that would be needed every day is then calculated as follows:

$$\text{Number of workers needed per day} = 40 \text{ craft hours per day}/8 \text{ hours per worker}$$
$$= 5$$

The five workers becomes the key control element in the input or performance management approach. The contractor would proceed to monitor the presence of the five workers each day. The first day that five workers are not present, it follows that the work will not be done in the budgeted 12.5 days.

This control approach would appear especially appropriate for monitoring a subcontractor's performance. By getting the contractor to set out how he is going to get something done, the contractor can then hold him to it. Instead of coming down on the firm when they fail to meet a promised completion date, the contractor would monitor the subcontractor's effort. The premise is simple, if the effort is not made, the results will not happen.

## 12.7   Record Keeping and Control

The job control reports discussed in this chapter are only as good as the data on the source documents that provide information to the reports. The source documents include the daily timecards, and daily or weekly reports that summarize the project quantities put in place.

Often the contractor's projects are quite distant from the firm's main office. In addition, many job site personnel may favor the production function over the field reporting function. Many contractors have difficulty getting time and accurate job site data to support the control function. In fact, inaccurate or late job site data often are characteristic of the entire construction industry.

Inaccurate or late field reports do not have to be part of the construction process. In fact, it is critical that the contractor obtain accurate and timely field data to include timecards and quantity reports. How can this be achieved?

Construction craft foremen and superintendents are often the individuals required to submit the data that serves the basis of the job control report. In requesting these individuals to fill out job forms, it is important for the contractor to remember and implement the following three rules:

- No one should fill out forms unless they are knowledgeable where the data goes, why it is needed, and how it is used.
- No one should fill out forms unless they are shown by example that the data is in fact used.
- No one should fill out forms unless they are given a feedback report (either written or oral) for every form they fill out.

The importance of accurate and timely record keeping should be emphasized with all workers. The contractor should consider giving recognition or even an award to individuals that fill out accurate and timely reports. Awards are often given for worker years of service, safety, etc. Given the importance of the control function, an award for accurate and timely record keeping seems equally important.

The very design of field reports affects the accuracy of the data on the field reports. For example, if the firm uses a weekly timecard, the foreman will likely procrastinate and only fill it out on the fifth day, unsuccessfully trying to recall what workers did two or three days ago. A daily timecard should be used. This will force daily recording of data.

The work codes on the timecard should be easily understood by the person filling out the form. Similarly forms that require field personnel to simply have to check items rather than write long narratives are preferred given the time pressures of the field personnel. The contractor should always follow the premise that field personnel are production people, not accountants. Anything that can be done to make the field reporting easier should be done.

## 12.8   Summary

The control function entails five steps: (1) the timely (2) comparison of the actual (3) to the plan (4) to detect a potential problem and (5) the follow up attempt to correct the problem. The control function entails the collection of accurate and timely job site information to include daily reports and time cards.

The control function is not limited to the control of project cost. The contractor must also implement procedures for controlling project time. By using common work items or cost objects, the contractor can integrate the project estimating, scheduling, and control systems.

In addition to project time and cost, the contractor needs to be attentive to controlling project quality and safety. The estimate and project plan and schedules set out the potential for project profits. It is the control function that enables the construction firm to realize profits.

| Work item | Budgeted quantity | Budgeted hours | Budgeted crew | Planed duration days | Actual quantity | Actual hours | Actual crew | Forecasted duration | % Work done | % hours | Forecast var. hours | Forecast var. time days |
|-----------|-------------------|----------------|---------------|----------------------|-----------------|--------------|-------------|---------------------|-------------|---------|---------------------|-------------------------|
| Form | 500 | 500 | 5 | 12.5 | 250 | 350 | 10 | 8.75 | 50 | 70 | +200 | −3.75 |
| Rebar | 200 | 300 | 6 | 6.25 | 100 | 150 | 5 | 7.5 | 50 | 50 | − | +1.25 |
| Place conc. | 100 | 200 | 4 | 6.25 | 50 | 80 | 4 | 5 | 50 | 40 | −40 | −1.25 |

**Figure 12.6.** Job control for project time and cost.

| Project performed | Quantity of suspended floor slabs | Crew size utilized | Man-hours expended | Man-hours/ square foot | Cumulative mh/sfca |
|---|---|---|---|---|---|
| School 101 | 10,000 sfca | 20 | 1,000 | 0.10 | 0.10 |
| Hospital 102 | 12,000 sfca | 12 | 1,050 | 0.0875 | 0.0932 |
| School 103 | 20,000 sfca | 15 | 1,600 | 0.08 | 0.0869 |
| Office 104 | 9,000 sfca | 13 | 800 | 0.0888 | 0.869 |
| School 105 | 20,500 sfca | 25 | 2,700 | 0.132 | 0.100 |
| School 106 | 20,000 sfca | 20 | 3,000 | 0.150 | 0.1109 |
| Totals | 91,500 sfca | | 10,150 | | |

**Figure 12.7.** Past productivity data.

# Chapter 13

# Prevention and Preparation of Construction Claims

## 13.1   Introduction

The contractor firm needs to perform their work with the attitude that everything is going to go well. However they need to prepare their paperwork with the idea that everything is going to go wrong to include involvement in a disputed change order, or a construction claim.

A construction claim can be defined as the following:

> A request for additional compensation by a contractor for added work they allege to have performed or for added costs associated with doing base bid work under conditions they did not expect and are not responsible for incurring.

Claims can become a time-consuming and costly part of many construction projects. Cost records and documentation of events can prevent a dispute and therefore a claim. This the best claim for two entities; i.e., the best claim is the one that doesn't evolve. While the prevention of claims is the cure-all for many of the woes of the construction industry, reality has it that claims often take place. It is at this point in time that the ability to properly prepare a claim becomes a critical process for the damaged entity.

Claims can get resolved in at least one of four ways:

1. Settled out of court
2. Arbitration
3. Litigation
4. Alternative methods (mediation, dispute boards, mini trials, etc.)

The majority of disputes likely get settled out of court. The contractor's ability to settle a dispute out of court or the firm's ability to win their dispute in arbitration or litigation is dependent in great part on their record keeping and their ability to quantify financial damages.

## 13.2  Reasons for Construction Claims

There are many reasons why a construction claim can evolve. For one, the complexity of projects have increased. This results in more difficulty in regard to the designer's ability to prepare contract documents that have only one possible interpretation. This increased project complexity also results in increased technological difficulty in regard to performing the work. In other words, claims have likely increased because of contractor management and technological difficulty in regard to performing the work.

A second reason for the increased number of construction claims relates to the high inflation that has characterized the construction industry and the material shortages that have characterized the building process during recent years. Rapidly increasing costs have resulted in increased emphasis on constructing a project in a specified time period. Delays, even if relatively short in duration, can jeopardize the favorable economics of a project. A single month's delay may equate to several thousand dollars of additional cost for a large project given inflated material, labor, and equipment costs. Given the occurrence of these delays, it is natural that one of the parties involved in the process will seek financial damages via a claim process.

Material shortages and delays in the delivery of materials to project sites are a common reason for project delays. During recent years, the construction industry has witnessed material shortages. At various times structural steel, concrete, plumbing, fixtures, and so forth, have been in short supply. Given these shortages or delays in procuring material, claims for time delays or increased costs have become common.

Another reason for the increased number of construction claims relates to the trend of the profitability of many construction firms. Published financial data and ratios indicate that the profitability (as measured by the profits divided by contract revenue) of many types of construction firms has been decreasing over recent years. There is evidence to indicate that a construction firm currently operates on a net profit margin (profit divided by revenue) of 2 to 3% of profit. Ten years ago this margin was closer to 6%.

Undoubtedly there are exceptions to the decreasing profit margins. Some firms have been successful in actually increasing their margins. However, the majority of firms have witnessed a decrease. There are several reasons cited for this negative trend. Increased governmental regulation and paperwork and increased competition are frequently cited. Others point to the industry's inability to increase their productivity. The productivity problems of the industry were discussed earlier.

The result has been that some firms have not been able to offset their increased operating costs with increased productivity. Given the contractor's inability to pass on all increased costs to the project owner, the contractor's margins decrease.

It might be argued that the number of claims that have evolved also reflect the fact that several new project delivery systems have evolved to include construction management (CM), design build, turnkey construction, and so forth. Some of these project delivery systems have evolved without proper and sufficient supporting contract documents, quantified procedures, etc. The result has been contract misinterpretation, overlaps in responsibilities, and so forth. These events promote the existence of claims.

## 13.3   Types of Claims

Construction claims can be classified as one of four types:

1. Delay Claims
2. Scope-of-Work Claims
3. Acceleration Claims
4. Change of Site Conditions

### Delay Claims

The delay claim often occurs when one contractor is delayed as a result of the performance of another contractor. For example, a midwestern mason contractor that may have planned on performing its work in July may be delayed until December due to the delay of another subcontractor. Given a significantly lower worker productivity in December (owing to December weather) than in July, the contractor likely has a cause for additional compensation due to a lack of productivity.

Another common delay claim often results because of material shortages or delays. For example, assume a project owner has agreed to provide the owner purchased material to its project as an owner purchased material. The contractor that is engaged to perform other mechanical work on the project may be engaged to install the material as part of its overall contract. Assume that because of a manufacturer strike or the owner's decision, the material is not procured according to the planned schedule and the mechanical contractor, therefore, has to staff the project an extended period of time. Given the fact that the contractor likely incurs additional on-site supervision costs, the firm may pursue additional compensation via a delay claim.

## Acceleration claims

An acceleration claim often results because a contractor is asked to perform its work in a period of time less than anticipated. For example, assume a project falls behind schedule for any number of reasons beyond the contractor's control. Assume that the project owner, desiring to get its project back on schedule, requires the contractor to work overtime. For example, the contractor may be required to work 6 ten-hour days instead of his planned forty-hour week. In addition to the added cost of the overtime premium, the contractor might argue that his labor productivity is adversely affected by the fact that the craftsmen are not as productive working a sixty-hour work week versus a forty-hour work week. In other words, a craftsmen will produce less work per hour when he works a sixty-hour work week than when he works a forty-hour work week. If the contractor can quantify the decreased productivity via the use of field and accounting reports, he can likely pursue an acceleration claim.

## Scope-of-work claims and changed-site-condition claims

Poorly written contract documents to include contract specifications are the leading cause for a scope-of-work claim. A contractor may have anticipated a certain quality of concrete finish work or degree of required project clean-up based on its interpretation of contract documents. However, the project owner's subsequent interpretation may indicate a different scope-of-work to be performed. If the contract documents can reasonably be interpreted to substantiate the contractor's interpretation, the contractor is entitled to make claim for additional compensation via a scope-of-work change claim.

Similar to the scope-of-work change claim is a claim resulting from a change of site conditions. Independent of the contract documents, a contractor is often confronted with changed site conditions when constructing a project. On occasion it can be argued that these conditions could not have been foreseen. For example, when excavating for a building foundation, a contractor might confront an unexpected layer of rock when driving piles. Assume that borings for the project did not indicate the presence of rock. No other evidence was available regarding the existence of the rock. Depending on the wording of the contract, the contractor might be successful in seeking additional compensation via a claim for a change in site conditions. Changed site condition claims are more prevalent in highway construction than they are in building construction.

Other types of claims may stem either from the suspension of work, a work stoppage, or a claim evolving from variations in unit quantities. These claims are less frequent in occurrence.

# 13.4    Warning Signs of Claim Situations and Documentation

Perhaps the best means a contractor, the project designer, or the project owner can use to prevent a claim is to be able to recognize warning signs of a potential claims. Claims seldom evolve without warning.

Almost every claim situation is unique. It is difficult to identify each and every specific situation that leads to a dispute or claim. Nonetheless, the following list represent circumstances or issues that may lead to the contractor's involvement in a claim situation.

| | |
|---|---|
| Time delays | Changed subsurface conditions |
| The need to accelerate work | Rock (quantity, hardness) |
| Interruptions | Weather changes |
| Unit price changes | Quality of workmanship changes |
| Quantity revisions | Unexpected test results |
| The need to work premium time | Ambiguities |
| Owner equipment delays | Defective drawings |
| Site access changes | Loss of efficiency |
| Drawing errors | Inspector change |
| Added moving time | Defective specifications |
| Escalation | Downtime |
| Voluntary work | Failure to approve |
| Unilateral change order | Out of sequence work |
| Emergency work | Out of season work |
| Terminations | Revised requirements |
| Denial of time extensions | Defective work |
| Deduction | Engineering changes |

Documentation of facts and figures is a key element to both the prevention and preparation of a claim. Oral interpretation is no substitute for recorded events. In a court of law, written evidence overrides oral testimony. More important, the fact that events are recorded will typically result in the prevention of a claim in that the recorded information will identify responsibility for an event and likely point to cause and effect. Documentation typically removes the question of "whose fault was it", "what was the delay", "why did it happen", etc.

One can argue that project documentation enables the contractor to be able to "play the project backwards like a movie" from the job site records. This includes the use of an effective daily report, time card, material receipts, purchase orders, billing statements, etc.

# 13.5    Types of Claim Cost Components and Types of Claims

Part of the ability to prevent or prepare a construction claim is dependent on the contractor's ability to recognize the type of claim damage that is part of a given type of claim. There are essentially four types of construction claims: (1) a scope-of-work claim, (2) a delay claim, (3) an acceleration claim, and (4) a changed-site-condition claim.

Figure 13.1 outlines the types of costs that might be considered relevant to any one of the four types of construction claims noted above.

| Type of cost claimed[a] | Delay claim | Scope of work claim | Accelera-tion claim | Changing site condition claim |
|---|---|---|---|---|
| Additional direct labor hours | | x | | x |
| Add'l direct labor hrs due to lost productivity | x | o | x | o |
| Increased labor rate | x | o | x | o |
| Additional material quantity | | x | o | o |
| Additional material unit price | x | x | o | o |
| Additional subcontractor work | | x | | o |
| Additional subcontractor cost | x | o | o | x |
| Equipment rental cost | o | x | x | x |
| Cost for owned-equipment use | x | x | o | x |
| Cost for increased owned-equipment rates | o | | o | o |
| Job overhead cost (variable) | o | x | o | x |
| Job overhead cost (fixed) | x | | | o |
| Company overhead costs (variable) | o | o | o | o |
| Company overhead costs (fixed) | x | o | | o |
| Interest or finance cost | x | o | o | o |
| Profit | o | x | o | x |
| Loss of opportunity profit | o | o | o | o |

**Figure 13.1.** Type of cost claimed.

[a]'x' indicates that the cost or component is normally included in the claim damage request. 'o' indicates that the cost or component may or may not be included in the claim damage request. The cost or component is normally subject to considerable dispute between the contractor and the entity he is claiming against (i.e., the project owner or designer). ' ' (blank) indicates that this type of cost or component is not normally included in the claim.

We will now describe briefly why each cost or claim component has been classified as being absent, almost always included, or subject to dispute in regard to a given type of claim. Because a changed site condition claim is similar to a scope of work claim, the components of a changed site condition claim will not be described.

## Delay Claims

In this section we focus on the types of claim damages that are commonly included in a delay type of construction claim. The intent is to make the contractor aware of the types of costs normally included in this type of claim, with the objective of being able to capture these costs and thus prevent or prepare a claim.

Before singling out the types of costs included in a delay claim, we should first characterize a delay claim. A delay claim normally results in a contractor's alleging that he has had to take longer to do required work than he had planned due to project owner interference, design changes, or project owner or designer indecision. Another possible delay claim may stem simply from the fact that the contractor was not able to start work when he had planned, and owing to a late start, ended up doing the work in a period (e.g., during cold weather) when his productivity was lower and thus the work took longer than he had planned. In either of these types of situations, the delay claim is characterized as resulting in the contractor's having to work longer than he had planned, independent of the fact that he still only performed the work he had originally planned and estimated.

### Additional direct labor hours — not included.

If one considers our additional direct labor hours category to include only those direct labor hours that are incurred due to the need to perform additional work not covered by the contract, this cost or claim component would not be part of a delay claim. No additional work is part of a delay claim. Instead, a delay claim relates to performing the same work over an extended period.

### Additional direct labor hours due to lost productivity — normally included.

Most delay claims are characterized as including a significant alleged lost-labor productivity claim. In fact, it is not unusual to witness a delay claim where the alleged lost labor productivity claim component is the largest dollar components in the overall claim.

The reason for including additional labor hours for lost productivity usually relates to the fact that the on-site labor crafts are delayed, have their productivity cycle broken, and may be required to stand by idle, owing to the alleged delays. Given delays, especially short delays that prevent a contractor from adjusting his on-site labor forces, the labor hours may not be as productive as initially planned. The difficulty associated with the additional direct labor hours owing to the lost productivity claim relates to the fact that it may be difficult for the contractor to quantify the lost productivity, owing to inadequate field records.

### Increased labor rate — normally included.

One of the easiest delay claim costs to quantify and usually prove is the increased labor rate associated with the payment of on-site labor during the extended period. If labor rates are adjusted upward (e.g., due to a labor union wage rate agreement), and a contractor is in fact delayed, it becomes a relatively straightforward process to multiply the number of craft hours incurred in the delay period by the increased labor rate.

**Additional material quantity — not included.**

A pure delay claim does not include additional work, only additional time. Thus a request for a cost for additional materials would not normally be included in a delay claim. Perhaps if the delay resulted in spoilage of materials on the job site, there might be some validity to including a cost for added materials.

**Additional material unit price — normally included.**

Similar to the claim for additional labor rates due to the extended period labor rate increase, a claim for the additional material prices paid for material purchased in the extended period would normally be included in the claim damage. However, the contractor would have to be able to show that there was indeed an increase in material prices and also show that he had no plans to purchase the material until it was needed (i.e., that he did not prepurchase the material to avoid the increased price).

**Additional subcontractor work — not included.**

Given the fact that a delay claim does not include additional work, no request for additional subcontractor work is part of the claim.

**Additional subcontractor cost — normally included.**

This is similar to the request for additional material prices paid in the extended delay period. If the contractor can show that he bought out his subcontractors as needed, and that subcontractors worked in the extended period and increased their quotes for inflation relative to the quotes they would have given prior to the time period, the additional cost would be part of the claim.

**Equipment rental cost — sometimes included.**

The contractor may attempt to show that during the delay period he had to remove his equipment to another job that had been planned to construct. Given the fact that he did not expect the delay on the project in question, and given the need to use his equipment elsewhere, the contractor may allege that he had to rent equipment and incurred an unexpected rental cost.

**Cost for owned-equipment use — normally included.**

If the delay results in work being performed that requires the contractor's equipment during the extended period, the contractor will normally include a claim damage for the use of the equipment for the extended period. The difficulty with this claim component may be that the contractor does not keep detailed ownership costs and therefore may have difficulty quantifying the hourly or daily rate for the added equipment time.

### Cost for increased rent — equipment rates-sometimes included.

If the contractor can show that he had planned to rent equipment and the equipment was to be rented as needed, he may attempt to show that the equipment rental rates paid in the extended delay period exceeded the rates that would have been charged without the delay.

### Job overhead costs (variable) — sometimes included.

Job overhead costs include costs that might be considered variable in regard to work activity. One might argue that the contractor can vary these costs from period to period. For example, he might be able to transfer some on-site supervision personnel from the delayed job to another job while a delay is in process. If the delays are short in duration but frequent, the contractor has little opportunity to "manage" variable job overhead, and therefore this cost would be included in the claim. However, if the delay consists of a single relatively long delay, the project owner may take the position that the variable overhead should have been reduced during the delay period and therefore not included in the claim.

### Job overhead costs (fixed) — normally included.

Unlike variable-cost job overhead, some job overhead costs, such as the rental of a job-site trailer, are fixed in nature in that the cost is incurred independent of work activity. These types of job overhead costs are incurred proportional to project duration.

### Company overhead costs (variable) — sometimes included.

Most company overhead costs are fixed in nature and occur mainly as a function of time. However, to the extent that a company overhead cost can be "managed" (e.g., office personnel time to travel to a job), one might argue the validity of including this portion of company overhead in the delay claim.

### Company overhead costs (fixed) — normally included.

This cost, including home office salaries, accounting expenses, and so on, is usually allocated to the delay claim via the use of a formula (e.g., the Eichleay formula). Often, this claim component is second in dollar amount only to the alleged cost due to lost labor productivity.

There often is dispute in regard to the amount of overhead cost to be included in any claim, including a delay claim. Given the allocation process by which the amount is determined, it is difficult to prove the actual amount incurred via accounting reports.

**Interest or finance costs — normally included.**

Because it is usually easy to single out the extended period involved in a delay claim, it is a relatively simple process to calculate a lost interest or finance cost associated with receiving progress payments later than originally planned or interest related to the investment of planned profit. This cost is usually included in a delay claim. Contract laws, however, may prevent payment of interest to the contractor.

**Profit — sometimes included.**

After the contractor has added all the claim cost components noted above, he may add or mark up the costs for additional profits. This "added profit" on claim items is subject to considerable dispute by the project owner in that he may question why the contractor should "gain a profit" due to an unfortunate claim event, a profit that he would not have earned otherwise.

**Loss of opportunity profit — sometimes included.**

A claim component that may be included but is subject to much speculation is a loss of opportunity profit associated with being at the project longer than anticipated. The contractor may take the position that, if his resources had not been tied up on the job in question, they could have been earning a profit on another job. The difficulty here relates to the ability to prove the opportunity profit amount.

## Scope-of Work Claims

In general, scope-of-work claims entail the contractor's alleging that he had to do work outside the original contract, or had to do work that was covered under the original or base bid contract but which had to be done under work conditions that he could not have forecast or included in his bid. Given adverse unexpected conditions, the contractor alleges that the cost of performing the work was greater than he expected-thus the scope-of-work claim.

**Additional direct labor hours — normally included.**

A scope-of-work claim entails an allegation that additional work was performed or that the work was made more difficult owing to unforeseen con-

274 Chapter 13. Prevention and Preparation of Construction Claims

ditions that are the responsibility of the project owner or designer. Thus a significant portion of the overall scope-of-work claim is usually the request for additional cost related to the additional direct labor hours incurred to do the work.

Although the direct-labor-hour component is an obvious part of a scope-of-work claim, the quantification of the claim amount is made difficult by the fact that the contractor's accounting records may not segment the additional hours from the hours required to perform the base bid work. As such, there is considerable dispute as to the amount, even if there is justification for the type of cost.

**Additional direct labor hours due to lost productivity — sometimes included.**

Given the existence of more work than anticipated, or the upsetting of work that was anticipated, the contractor may make a claim to the fact that the "normal" or planned work was also affected and that this work required more labor hours than anticipated. He uses a cause-and-effect argument. For example, he may allege that he had to "break" his ideal crew size for normal work in order to staff the work in question under the scope-of-work claim. In effect, he puts in a claim for more hours than merely those directly related to the scope of work under question.

The difficulty with the allegation for hours due to lost productivity relates to being able to prove this cause-and-effect relationship and being able to quantify the exact dollar impact.

**Increased labor rate — sometimes included.**

Normally, an increased labor rate would not be part of a scope-of-work claim in that the scope-of-work claim does not necessarily imply an extended project duration that may result in a wage differential. However, if the added work requires the contractor to work overtime, or does in fact result in a project time extension, the increased labor rate cost could conceivably be part of the scope-of-work claim.

**Additional material quantity — normally included.**

Another obvious scope-of-work claim component would be the cost of purchasing the added material for the alleged work in question. As with the additional direct labor hours, there may be some difficulty in segmenting the additional material cost from the cost that is included in the base bid contract.

**Additional material unit price — not included.**

Given the assumption that the additional material quantities needed to perform the work in question are included in the scope-of-work claim, there is no need to include a figure for additional costs for material price increases. This statement is made with the assumption that the actual prices paid are included in the material quantity claim request.

**Additional subcontractor work — normally included.**

To the extent that the scope-of-work claim includes work performed by subcontractors, the subcontractor cost would be included in the contractor's claim.

**Additional subcontractor cost — sometimes included.**

Given the assumption that any additional work performed by subcontractors is included in the additional subcontractor work claim component noted above, normally an additional subcontractor cost would not be included in the claim. However, if the additional work performed by the contractor upsets the subcontractor's work or requires him to perform the work later in the schedule or requires him more time to perform his work, the claim may include an additional subcontractor cost for time delays or costs related to inflation.

**Equipment rental cost — normally included.**

This cost is usually a logical part of a scope-of-work claim in that the contractor can claim that due to the additional work, he had to procure additional rental equipment over and above his owned equipment to perform the work in question. The cost is usually relatively easy to quantify.

**Cost for owned-equipment use — normally included.**

Should the additional alleged work require the use of equipment, and the contractor is required to keep his job-site equipment on the job longer or finds it necessary to send more of his owned equipment to the job site to perform the work, it is logical that an additional cost for owned equipment be included in the scope-of-work claim. As is true of all owned-equipment-cost claim components, a difficulty results because of the contractor's inability to quantify the claim amount from accounting records.

**Cost for increased owned-equipment rates — not included.**

Unless the additional work results in a distinct time period extension, there usually is little basis for including a cost for planned versus actual equipment rates.

**Job overhead costs (variable) — normally included.**

To the extent that additional job overhead costs are required to support the alleged additional work, there is justification for including these types of job overhead costs in a scope-of-work claim. For example, if additional job supervision is needed to oversee the additional work, this cost may be included in the claim.

**Job overhead costs (fixed) — not included.**

By definition of a fixed cost, such job overhead fixed costs as trailer rent or monthly on-site utility costs occur mainly as a function of time and may have little relationship to the amount of work performed in the specific period. Thus they normally would not be included in a scope-of-work claim.

**Company overhead costs (variable) — sometimes included.**

The contractor takes the position that company overhead costs are incurred to support construction work the firm performs. As such, given additional work that has to be performed, the contractor uses an allocation process to apportion some of his company overhead to the claim request. There is probably a strong argument for this allocation for the variable overhead costs he incurs in that he can argue that the additional work requires additional amounts of the variable overhead costs.

**Company overhead costs (fixed) — sometimes included.**

The same argument for including these fixed overhead costs can be made as the one for the variable fixed overhead costs. However, given the fact that the term fixed implies independent of activity, this somewhat weakens the argument for including the fixed-cost component of company overhead.

**Interest or finance costs — sometimes included.**

Given the fact that the scope-of-work claim may not necessarily imply an increase in project time, an interest or finance cost may not be a common scope-of-work claim component. However, to the extent that payment is not made on the alleged scope-of-work claim addition, the interest on this payment may be part of the claim.

**Profit — normally included.**

If one views profit as the return for performing services, it may be logical to include additional profit on the scope-of-work change in the claim-the only question is: how much? Nonetheless, there appears to be a stronger case for including profit in a scope-of-work claim than in a delay claim.

**Loss of opportunity profit — sometimes included.**

If the additional work causes the contractor to utilize resources that normally would have been used on another project, he may include a loss of opportunity profit associated with "tying up" his resources on the additional work. The difficulty relates to proving or quantifying this lost opportunity profit.

## Acceleration Claims

Before singling out the types of costs normally included in an acceleration claim, it is important to characterize an acceleration claim. The construction process is characterized by a group of interdependent resources working together at the job site. For example, one construction craft may have to wait for another craft. A subcontractor may be delayed because of the non-performance of another contractor. A contractor may have to wait for the delivery of material from a supplier. The result is that the production process may be slowed or brought to a halt because of the non-performance of another resource.

Given this interdependence of resources, an acceleration claim may result. For example, if a single contractor is delayed in his performance, the project owner may require a subsequently performing contractor to "hurry up" so that the project owner can finish the project on schedule.

This hurrying up of the construction process can result in the contractor's incurring more costs than the firm included in the original estimate. This may result because of the need to use a larger crew size, having to work additional hours, using additional equipment at the job site, and so on. Any one of these work changes can result in the contractor's incurring unexpected and additional costs.

An acceleration claim is characterized by the contractor's having to use additional resources during a specific time period or having to perform planned work in a shorter period than the contractor had originally planned. Given this characteristic, the following types of costs are commonly included in the acceleration claim.

## Additional direct labor hours — not included.

If one considers additional direct labor hours to include only those direct labor hours that are incurred to perform additional work not covered by the initial contract, this cost or claim component would not be part of a delay claim. No additional hours are necessarily incurred because of a need for the contractor to "hurry up" the production process. The same work is performed. Thus, an acceleration claim does not normally include a claim damage for additional direct labor hours.

## Additional direct labor hours due to lost productivity — normally included.

Perhaps the largest cost claim component usually included in an acceleration claim is the alleged additional direct labor hours due to lost productivity. Two of the most common types of acceleration claims relate to the contractor's having to use larger-than-planned crew sizes or having to work his original planned crews for additional work hours in a day or week (i.e., overtime). For either of these events, the contractor commonly alleges a negative impact on his direct labor productivity and cost.

The principle behind the alleged negative damage on labor productivity relates to the fact that there is an ideal crew size for every work activity that the contractor must perform. In addition, there is an ideal workday for which optimal productivity can occur; a 5-day work week including 8-hour days.

For example, in placing concrete wall forms a contractor may decide that his ideal crew size for optimizing labor productivity is 5 carpenters and 2 laborers. Based on his experience in performing the work in question, he has determined that any crew size bigger or smaller than this "ideal" crew size results in less work performed (i.e., contact area of wall forms erected) per labor hour of work input.

Given the assumption that the information above is in fact true and given instructions from the project owner or his designer to increase the crew size to speed up the production process (perhaps because a prior contractor fell behind schedule), the contractor may allege a negative impact on his productivity.

Similarly, a contractor may be required to work his so-called "ideal crew" more hours than he had planned. For example, the contractor may be required to work six 10-hour days, owing to the need to catch up for a prior contractor's non-performance. The contractor will probably state that his productivity or efficiency is negatively affected due to the fact that any one of his workers' outputs is reduced relative to the output he would perform in a normal 8-hour 5-day work week. Many technical studies have been conducted to support a claimed loss of labor productivity when a worker works overtime.

Acceleration claims can also result because of the need to hurry up after delays resulting from an owner's failing to make a decision, or after a delay in purchasing or obtaining material. As is true of any alleged lost productivity time or hours claim, the difficulty of a "lost productivity" acceleration claim is the problem of quantifying the alleged lost cost. The contractor may not have segmented costs in a way that enables separation of the alleged "lost labor productivity" cost from the labor cost included in the original bid estimate.

**Increased labor rate — normally included.**

An increased labor rate claim component would obviously be part of an acceleration claim that resulted due to overtime if the overtime required an increased hourly wage rate. It is common for a worker to be paid time and a half or double time if the worker is required to work more than 8 hours on a given day or more than 40 hours in a given week.

Other than an overtime acceleration claim, there usually is not a valid reason for including a labor rate claim component in an acceleration claim.

**Additional material quantity — sometimes included.**

One would not normally consider a need for added material to result from an acceleration claim. However, on occasion, a contractor has included this type of cost in his overall acceleration claim request. The reason may stem from the fact that in a need to hurry up the production process, the contractor may allege that he had to handle material in a rushed fashion, the result being some material breakage or excessive waste. This may be difficult to prove. The project owner is likely to take the position that any additional material needed stems from the contractor's mishandling of the material and not an acceleration process.

**Additional material unit price — sometimes included.**

An acceleration claim usually implies a "hurry-up" situation, a situation normally resulting in a contractor's completing his planned work in a shorter period than originally planned. Thus an increase in material costs is not normally included in an acceleration claim in that the late purchase of material that would imply an inflation cost would not occur. However, on occasion, the contractor may allege that due to the acceleration, he had to buy material at a non-optimal time relevant to when he had planned. The end result is that the contractor may include an increased material rate in his claim, though only rarely.

**Additional subcontractor work — not included.**

Given the fact that an acceleration claim does not include additional work, there is usually no request for additional subcontractor cost to be included in such a claim.

**Additional subcontractor cost — sometimes included.**

This is similar to the request for additional cost for an increase in material rates. The contractor may imply that a subcontractor has to perform work in a non-optimal or planned period that results in the subcontractor's incurring additional costs. However, this is rare, and a request for additional costs related to subcontractors' incurring added costs is not normally included in a claim.

**Equipment rental cost — normally included.**

Given the fact that an acceleration claim usually implies a speeded-up production process, a contractor may allege that he had to rent additional equipment to complete the work in the accelerated period. This additional rental equipment may be over and above the rental equipment and/or owned equipment he had planned to use. Thus costs related to equipment rental are normally included in an acceleration claim.

**Cost for owned-equipment use — sometimes included.**

Given the fact that an acceleration claim is not accompanied by additional work, the contractor cannot allege that additional owned equipment is needed to perform additional work. However, to the extent that the speeded-up production process results in the contractor's having to use more of his owned equipment than he had planned, the hourly ownership cost of this equipment may be included in the claim.

**Cost for increased owned-equipment rates — sometimes included.**

If the contractor can show that he planned to use his equipment, but due to the need to accelerate the construction, had to rent equipment, he may attempt to seek the additional rental rate for rental equipment that is over and above his rate for owned equipment.

**Job overhead costs (variable) — sometimes included.**

Variable job overhead costs occur as a function of work activity. The more work activity, the more variable job overhead costs, such as supervision, are required. Given the fact that an acceleration claim does not involve any more work than additionally planned, one might argue that variable job overhead costs should not be included in an acceleration claim. However, the contractor may take the position that given the need to hurry the work process, more costly variable job overhead had to be procured than originally planned. For example, given the hurried-up production process, the contractor may find it necessary to engage a second superintendent to support the originally planned work. This additional superintendent may be at a rate over and above the cost that would have been charged had the planned superintendent supervised the work in the initially planned time period.

### Job overhead costs (fixed) — not included.

By definition, fixed job overhead costs such as trailer rental occur as a function of time. Thus they are not affected by the amount of work done in a specific period. It follows that the speeded-up work process that characterizes an acceleration claim should not add additional fixed job overhead costs.

### Company overhead costs (variable) — sometimes included.

Few company overhead or home office costs are variable in nature. However, some may be considered variable in part, such as home office accounting costs that are incurred to support work activity. Given the fact that a firm does not incur many variable company overhead costs, this type of cost is not normally included in an acceleration claim. However, to the extent that the contractor can show the occurrence of additional variable company overhead costs, such as additional accounting costs, the contractor may attempt to include these types of costs in an acceleration claim.

### Company overhead costs (fixed) — not included.

Given the fact that fixed company overhead costs are independent of work activity, and given the fact that an acceleration claim is characterized by a speed-up of the production process, there is little basis for including a fixed company overhead cost in an acceleration claim.

### Interest or finance costs — sometimes included.

Given the fact that an acceleration claim results in a distortion of the original planned construction schedule and/or a hurried-up production process, the contractor may allege that additional financing is required to perform the accelerated work. For example, it may be that the need to incur and pay for more craft hours than planned may require the contractor to obtain additional working capital that can be obtained only by more financing — financing that results in additional interest costs.

**Profit — sometimes included.**

Given the fact that no additional work characterizes an acceleration claim, one might suggest that no additional profit is justified. However, it is common for the contractor to add a certain percentage profit on the other costs included in his claim. This is done with the principle that profit is added to other costs in preparing the original bid estimate. Therefore, if one considers profit that one earns for incurring costs, one might argue that profit can be "added" as a percentage of other claim costs.

**Loss of opportunity profit — sometimes included.**

A somewhat speculative cost, but a cost that may be included in an acceleration claim, is a loss of opportunity profit. For example, in having a need to accelerate work, the contractor may allege that he had to take an additional superintendent off another project or use equipment that he had planned to use on a different project. The contractor may allege that either of these events may result in a negative impact on the planned profit of the other project. Thus he may attempt to quantify this "lost" profit as part of the acceleration claim. The difficulty relates to trying to quantify the alleged lost profit.

# 13.6   Quantification of Three Primary Claim Cost Components

The above section has illustrated the many cost components that may be part of the quantification of a disputed change order or claim. However, experience has shown that when the contractor has a disputed change order or a claim, that the following three components normally account for as much as ninety to ninety-five percent of the total cost.

1. Lost labor productivity cost (commonly referred to as impact)
2. Cost of using owned equipment
3. Company overhead extension cost

   Given the importance of these claim components, it is important that the contractor understand how to quantify these components.

## Methods of Quantifying Lost Labor Productivity

### Discrete Method

In the discrete method, the construction job site personnel attempt to record the alleged productivity inefficiencies directly on the daily time card

| Description | Hours |
|---|---|
| Erect slab forms | 2.0 |
| Place rebar | 2.5 |
| Place concrete in slab | 2.5 |
| Waiting on materials | 0.5 |
| Waiting on drawings | 0.5 |
| Total | 8.0 |

**Figure 13.2.** Daily timecard — discrete method

or daily report. "Impact" codes are created and lost time is recorded as shown in Figure 13.2.

This approach is often viewed as the best approach in that the lost hours are identified and segmented on the time card. In reality, this method is seldom used. For one, experience has shown that many contractors don't get serious about a problem until it is near the end or over. This prevents them from recognizing the hours on the time card as they occur. Secondly, time cards are traditionally used to record labor hours against the budgeted hours. Because defects or time delays are not part of the budget, they are not normally recorded on the time card.

**Measured Mile Approach**

In this method, the productivity achieved during the alleged troublesome period is compared to the productivity achieved when there was not trouble or impact. This approach is illustrated in Figure 13.3.

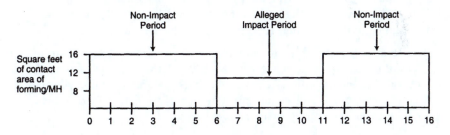

**Figure 13.3.** Measured mile approach.

The measured mile presents a graphical or mathematical means of illustrating alleged lost labor productivity. The method assumes that the work performed in the non-impact periods and the work performed in the alleged impact period were similar.

The measured mile exhibit can be prepared as the claim issue evolves or can be prepared from past records. Therefore this approach can be prepared after the fact.

## Total Cost Method

In this method, the alleged labor productivity cost damage is calculated as the difference between the actual project labor cost or hours and the estimated labor cost or hours. An example is illustrated in Figure 13.4.

| | |
|---|---:|
| Actual Project Labor Hours to place | |
| 1,000 cubic yards of concrete | 650.0 mh |
| Estimated Project Labor Hours to | |
| Place 1,000 cy of concrete | 480.0 mh |
| Difference | 170.0 mh |
| times Project Labor Rate | $30.00/hr |
| Total Claim Damage | $5,100.00 |

**Figure 13.4.** Total cost method.

In affect the total cost method assumes the following:

1. The method assumes that the estimating process was correct and that the resulting estimated hours and cost amount is correct.
2. The method assumes that the recorded project labor cost and hours is correct.
3. The method assumes that all the reasons for the overrun in the labor cost damage are the fault of the project owner or designer.
4. The method assumes that the labor rate used in the calculation is a fair and equitable rate and represents what the contractor paid.

The total cost method is the most frequently used method for quantifying lost productivity. This is likely because the method can be used after the fact, even when the project is complete. Given the fact that many firms don't get serious about a problem until late in the project, it follows that the total cost method is frequently utilized.

## Industrial Engineering Approach

In this approach, industrial engineering models such as time study and work sampling is used to measure productivity impact as it occurs. While this approach represents the most scientific method, it is seldom used. For one, the approach requires the contractor to be knowledgeable about various industrial engineering models. Secondly, the method or approach assumes that the contractor documents the problem as it happens.

**Earned Value Method**

In the earned value approach, the calculation of added labor hours is determined by comparing (for select time periods), the hours it took to do a group of activities versus the budgeted (or earned hours) for the activities. To illustrate the Earned Value Method, consider the sample small project described in Figure 13.5 that represents the planned work durations, planned work days, and planned man-hours relative to the actual durations, and work days.

Assume the schedule per the contractor's submitted "as planned schedule" was as shown in Figure 13.6.

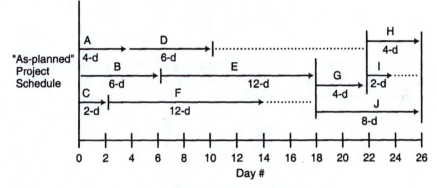

**Figure 13.6.** As planned schedule.

Let us further assume that for reasons outside the control and responsibility of the contractor, that labor productivity impact occurred during days 11 through 20. Let us further assume that for whatever reason, project activities are delayed and shifted and the "as built" project schedule for the above illustrated project is as illustrated in Figure 13.7.

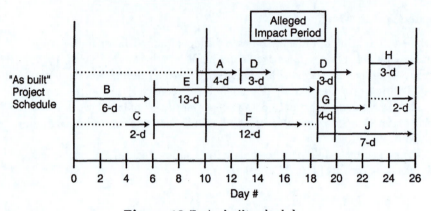

**Figure 13.7.** As built schedule.

Assuming the contractor has kept track of the actual man-hours expended every day (from the timecards), the lost productivity or man-hours calculated using the Earned Value Method is as shown in Figure 13.8.

| Day | Actual hours | Earned hours | Lost hours (prod.) |
|-----|--------------|--------------|--------------------|
| 1 | 32 | 48 | (16) |
| 2 | 32 | 48 | (16) |
| 3 | 32 | 48 | (16) |
| 4 | 32 | 48 | (16) |
| 5 | 62 | 80 | (18) |
| 6 | 62 | 80 | (18) |
| 7 | 60 | 53.5 | 6.5 |
| 8 | 60 | 53.5 | 6.5 |
| 9 | 60 | 53.5 | 6.5 |
| 10 | 93 | 85.5 | 7.5 |
| 11 | 93 | 85.5 | 7.5 |
| 12 | 93 | 85.5 | 7.5 |
| 13 | 93 | 85.5 | 7.5 |
| 14 | 108.2 | 93.5 | 14.7 |
| 15 | 108.2 | 93.5 | 14.7 |
| 16 | 108.2 | 93.5 | 14.7 |
| 17 | 60 | 53.5 | 6.5 |
| 18 | 60 | 53.5 | 6.5 |
| 19 | 81 | 69.5 | 11.5 |
| 20 | 126.8 | 116.6 | 10.2 |
| 21 | 126.8 | 116.6 | 10.2 |
| 22 | 78.5 | 76.6 | 1.9 |
| 23 | 78.5 | 76.6 | 1.9 |
| 24 | 100 | 100.6 | (0.6) |
| 25 | 159 | 164.6 | (5.6) |
| 26 | 159 | 164.6 | (5.6) |
| Total | 2158 | 2128 | |

**Figure 13.8.** EVM calculations.

The "Earned man-hours" for each time period are calculated through a linear interpolation of the time in the impact period. For example, for activity B, the planned or estimated man-hours was 288 man-hours. The actual duration for activity B was six days. Therefore each day when B was performed or worked, 48 hours are earned (i.e., 288/6). The "Actual Man-hours" for each time period would come from the time cards for the time period.

The calculations above for the sample project result in a determination 101.3 lost man-hours (productivity) owing to the impact issues. Note that this is higher than the total man-hour overrun of 46 man-hours for the project. The reason is that in time periods outside the "impact period", the contractor actually beat there budget; i.e., had a gain in productivity. As such, the analysis for the sample project shown above indicates that the contractor should be reimbursed for the 101.3 lost man-hours (productivity) that incurred during the impact period.

**Expert Approach**

In this approach, a productivity expert(s) review the project conditions, project correspondence, and project documentation, and applies industry productivity knowledge and data to the alleged issues at hand to determine a reasonable lost labor productivity damage. For example, the expert may reference the information in Figure 13.9 as presenting industry data regarding productivity loss as a function of various impact factors.

Obviously, when using the expert approach, the calculations and the resulting calculated labor cost damages are only as creditable as the individual expert that has reviewed the issues and documents and made the calculations.

## Quantifying Owned-Equipment Damage

An issue in dispute in claim resolution as well as for determining change order or claim payment is the issue as to how much the contractor is entitled to for his owned equipment when added work has been done or when the work has been delayed. This issue is made critical by the fact that the hourly cost for several types of equipment may approach $100 per hour or more.

Two separate situations may initiate this issue of owned equipment costs. For one, the contractor may have had to do added work; i.e., change order work. The project owner may take the position as long as the project took the same duration as what was planned that the contractor is not entitled to added costs for his owned equipment. The owner's position is predicated on the assumption that the owned equipment costs is a sunk cost, it was there anyway.

The second situation relates to a delay in a project that required the contractor to have the equipment at the project longer than anticipated. The owner may take the position that unless the contractor can show that there was another project waiting the availability of the equipment, that the equipment is a sunk cost and therefore the contractor is not entitled to added compensation.

To resolve the above issues, both the project owner and the contractor have argued for cost reimbursement for owned equipment using one of the following rates:

- Rental rates as published by industry guide estimating books (referred to as blue book rates)
- Sunk cost (owner position that equipment sunk cost and therefore contractor entitled to $0)
- Estimating rates
- Actual ownership costs

The actual rates are the most appropriate and equitable rate. Both of the above issues (added work or added duration) can be resolved if the contractor keeps track of the ownership costs of owning a specific piece of equipment. The costs of ownership include the following:

$$\begin{aligned} \text{Equipment cost} = \ &\text{Depreciation} \\ &+ \text{Maintenance} \\ &+ \text{Operating} \\ &+ \text{Repairs} \\ &+ \text{Finance cost} \\ &+ \text{Insurance cost} \\ &+ \text{Property tax} \\ &+ \text{Replacement cost (inflation)} \end{aligned}$$

Some the above costs are a function of equipment use:

- Depreciation (Note: Equipment wears out more as a function of use than of time.)
- Maintenance
- Operating
- Repairs

Other cost components are more a function of time:

- Finance cost
- Insurance cost
- Property tax (if there is a property tax component)
- Replacement cost

If the contractor knows each of the above cost components by having kept records, the contractor can justify requesting cost components that are a function of use when the contractor has an added work but not added time issue. Similarly, when the contractor is delayed with his equipment, even though the contractor did not do any added work, he can claim for the cost components that are a function of time.

## Quantifying Company Overhead Damage

When a contractor submits a claim or a change order for payment, he often includes a cost component for the firm's company overhead (company overhead including officer salaries, home office rent, etc.) The amount of this claim component is often large and disputed. Given the fact that most company overhead costs are a function of time (i.e., fixed costs), a company overhead cost component seems most appropriate to a delay claim. Nonetheless, it often included in most claims using one of the following procedures for quantification:

- Industry average rates as published in industry publications
- Company rate per financial statements
- Company overhead rate used in estimate
- Using the Eichleay Formula

The Eichleay formula is a process of allocating company overhead based on volume and time. Consider the example in Figure 13.10.

| | |
|---|---:|
| Project 101: | |
| Contract amount | $1,000,000 |
| Planned duration | 50 weeks |
| Delay (for reason outside control of contractor) | 5 weeks |
| Company volume in time period | $10,000,000 |
| Company overhead in time period | $500,000 |

**Figure 13.10.** Example information.

The contractor would calculate the company overhead damage (using Eichleay) as follows:

| | |
|---|---|
| Project 101 contract/Total company volume | $1,000,000/$10,000,000 = 10\% |
| Company overhead allocable to Project 100 | (.10) ($500,000) = $50,000 |
| Company overhead planned per week | $50,000/50 weeks = $1,000/week |
| Company overhead damage | (5 weeks) × $1,000/week = $5,000 |

The above can be considered the contractor version of Eichleay. Even if the owner agrees with the entitlement using Eichleay, the project owner will likely argue that the company overhead should be calculated by using the actual duration rather than the planned duration. This would result in dividing the $50,000 by 55 weeks instead of 50 weeks.

## 13.7   Summary

Construction disputes and claims result for many reasons; the most common reason being the very uncertain nature of the construction process. The contractor should construct a project with an attitude that everything is going to go right. However, the firm should do their paper work and field documentation with the idea that events may in fact go wrong.

Construction claims often result because of disputed change orders. Claims can be classified as either a scope-of-work claim, a delay claim, or an acceleration claim. To be successful in winning a dispute or a claim, the contractor must illustrate entitlement as well as quantify financial damages.

The three largest financial damages normally relate to the disputed lost labor productivity, the cost of using owned equipment at the job site, and the damages resulting due to extended company overhead. Often the largest and most difficult to prove damage is the alleged lost field labor productivity and labor hours. There are several approaches to quantifying this claim component. Included are the total cost method and the measured mile approach. Independent of the approach used, the contractor is likely to best be able to quantify damages if the firm becomes attentive to the dispute issues when they in fact occur. All too often, the firm becomes serious about problems too late. This places the firm at a disadvantage when it comes to proving entitlement and/or financial damages.

| Activity | Planned duration | Planned work days | Planned man-hours | Planned mh/day & crew | Actual duration | Actual work days | Actual man-hours | Actual mh/day & crew |
|---|---|---|---|---|---|---|---|---|
| A | 4 | 1–4 | 128 | 32–4 | 4 | 10–13 | 148 | 37–4.6 |
| B | 6 | 1–6 | 288 | 48–6 | 6 | 1–6 | 192 | 32–4 |
| C | 2 | 1–2 | 64 | 32–4 | 2 | 5–6 | 60 | 30–3.75 |
| D | 6 | 5–10 | 240 | 40–5 | 6 | 14–16,18–20 | 290 | 48.3–6+ |
| E | 12 | 7–18 | 384 | 32–4 | 13 | 7–19 | 424 | 32.6–4+ |
| F | 12 | 3–14 | 288 | 24–3 | 12 | 7–18 | 328 | 27.3–3.4 |
| G | 4 | 19–23 | 160 | 40–5 | 4 | 20–23 | 154 | 38.5–4.8 |
| H | 4 | 23–26 | 192 | 48–6 | 3 | 24–26 | 180 | 60–7.5 |
| I | 2 | 23–24 | 128 | 64–8 | 2 | 25–26 | 118 | 59–7.4 |
| J | 8 | 23–26 | 256 | 32–4 | 7 | 20–26 | 280 | 40–5 |
| Total | | | 2,128 | | | | 2,174 | |

Figure 13.5. Plan versus actual.

| Description | Range of impact on productivity per industry publications | Severity on example project | Estimated impact on example project | On-site labor hours during relevant time frame | Calculation of expected lost labor hours |
|---|---|---|---|---|---|
| Weather — temperature | (10–30%) | small | 10% | 820 | 82.0 |
| Inadequate scheduling | (5–15%) | severe | 15% | 940 | 141.0 |
| Inadequate supervision | (10–25%) | average | 16% | 244 | 39.0 |
| Worker morale | (5–30%) | significant | 25% | 128 | 32.0 |
| Interference by other trades | (10–30%) | small | 10% | 540 | 54.0 |
| Inadequate skills of workers | (10–30%) | significant | 28% | 98 | 27.4 |

**Figure 13.9.** Range of impact.

# Checklists for Evaluating Project Management Business Practices

The chapters in this book have set out both business practices for successful contracting to include the following functions:

- Estimating
- Planning and Scheduling
- Productivity Measurement and Improvement
- Project Control
- Claims Prevention and Preparation

The contracting firm should use the enclosed checklists in the following way. Place a check next to a suggested practice if the firm is currently not doing the practice. Take actions to implement the practice. Continue to use the checklist and identify weaknesses with the objective of seeking on-going improvement.

The concept is simple — fix the weakness, and the firm will be rewarded with increased profitability and will be able to construct projects on time, on budget, with acceptable quality, and enhanced safety.

# Appendix A

# Estimating Procedures

## Estimating Step/Process

**Firm Evaluated/Name** _____ **Date** _____

**Review of Contract Documents**

1. Supplementary conditions are reviewed or analyzed to determine:
   a. Scheduling requirements
   b. Requirements regarding contractor drawing submittal requirement to include installation drawings.
   c. Notification clauses regarding need to notify project owner or designer of problems or conflicts.
2. The firm thoroughly reviews contract drawings and specifications to determine work requirements, work difficulty, and possible drawing and/or specification errors and omissions.
3. A visit and review of the job site is made prior to the preparation of the estimate and bid to determine degree of work difficulty and job conditions that affect the cost of doing project work.

**Quantity Take-Off**

1. The firm utilizes checks and balances on take-off procedures and take-off math; e.g., cross footing of numbers on pages, math checks on mechanical take-off devices, etc.
2. The firm utilizes material checklists to ensure that all materials and work is properly taken-off.
3. The firm breaks out the take-off of materials into enough detail to ensure work can be properly costed.

## Determining Craft Hours and Labor Hours

1. The firm uses a data base of past productivity data to enable estimate of required labor hours.
2. The firm recognizes all work tasks that must be performed that relate to material installation; e.g., erecting scaffolding, type of concrete finish required, etc.
3. The firm utilizes correct labor rates for workers that are to perform construction work to include:
   a. Use of correct wage rates
   b. Use of correct crew mix; e.g., apprenticeships to journeymen that would effect the average wage rate.
   c. Use of correct labor burden rates.
4. The firm recognizes impact of when work will be done in regard to calculation of labor hours and/or labor rates.
5. The firm recognizes relationship of project schedule of work activities to the labor hours and labor cost required.
   a. The firm recognizes impact of multiple activities going on at the same time on labor productivity.
   b. The firm recognizes how many locations of work are to be going on at the same time which will result in a need for added supervision that affects labor productivity.
6. The firm coordinates estimate assumptions with supervisory personnel that will construct the project; i.e., the estimator may assume one method of doing a work task and an accompanying labor productivity, only to have the job site personnel utilize a different method.
7. The firm recognizes specific contract provisions regarding labor rates or burdens specific to a project; e.g., requirement to pay federally dictated rates on a specific project.
8. The firm recognizes specific labor work rules relevant to a specific location and/or project.

## Material Quantities and Costs

1. The firm obtains accurate quotes on materials needed for a project.
2. The firm considers when materials are required as a function of time which would influence the actual purchase cost of the materials.
3. The firm calculates and analyzes material wastage quantities.
4. The firm includes support materials such as related rebar, form ties, etc.

## Equipment Hours and Costs

1. The firm utilizes a formalized process to identify specific types of equipment needed for the project as a function of project time.
2. The firm includes schedule analysis to determine equipment needs and costs as a function of schedule needs.
3. The firm utilizes a checklist of established equipment rental rates for equipment that is rented for the job use.
4. The firm obtains owned equipment rates determined as a basis of determining equipment costs for a specific project.
5. The firm performs a formalized analysis of equipment needs as a function of work difficulty and as a basis of determining equipment hours needed.
6. The firm performs an analysis of equipment operating costs and maintenance costs for a project as a function of work difficulty and expected wear and tear.

## Subcontractor Costs

1. The firm utilizes a checklist as a reminder that all work that needs to be subcontracted has been included in the bid.
2. The firm uses a formalized process or checklists as a means of verifying and validating subcontractor bids.
3. The firm performs an analysis to evaluate the economics of doing self-performed work versus subcontracting work.
4. The firm uses a formalized process or checklist to determine special subcontractor bidding requirements; for example, small business set asides or minority requirements.

## Job Site Overhead and General Conditions

1. The firm utilizes a checklist as a reminder to identify all job overhead costs that will be needed for a project.
2. The firm utilizes and analyzes the project schedule to determine the needed duration of job overhead items to enable the accurate costing of the items.
3. The firm analyzes the work process to include the number of work tasks that are performed simultaneously in order to establish the amount of supervision that is required as a function of time.
4. The firm reviews contract documents to enable the identification of contract administration requirements that will add to the cost of job overhead and must be included in the estimate; scheduling requirements, submittal requirements, etc.

5. The firm does not determine job overhead costs as a percentage of another job cost rather than itemize and analyze specific job overhead items.
6. The firm does not determine job overhead items as company overhead or a part of the project markup instead of identifying and analyzing each job overhead cost item.
7. The firm does not use an allocation process for determining a cost estimate for small tools and supplies instead of performing an analysis of the costs.

## Company Overhead/Home Office Costs

1. The firm utilizes a formalized process for determining company overhead requirements as a function of project duration.
2. The firm recognizes project duration and project schedule as considerations in determining company overhead or home office costs needed to support a project.
3. The firm recognizes the required number of on-site labor hours that will be needed for a project as a consideration in determining company overhead or home office costs needed to support a project.
4. The firm considers annual company overhead of the firm as a factor in allocating company overhead to be burdened to a specific project.

## Contract Markup and Profit

1. The firm uses a formal process to recognize project risk as a consideration in determining profit to be included in the bid.
2. The firm considers project duration and project complexity as factors in determining profit to be included in bid.
3. The firm recognizes form of contract as a profit consideration.
4. The firm considers the number of expected bidders in determining profit to be included in bid.

# Appendix B

# Planning and Scheduling Procedures

## Planning and Scheduling Procedures

**Firm Evaluated/Name** _____ **Date** _____

**Timing of Schedule Preparation**

1. Schedule is not prepared independent of estimate.
   a. Input is obtained from job site superintendent of field personnel.
   b. Project duration and job site overhead expenditures are taken into consideration.
   c. Schedule of work activities reflects the timing of when activities are performed.
2. Schedule is prepared prior to project bid.
3. Timing of preparation of schedule is compatible with contract requirements
4. Schedule is not prepared independent of input from all subcontractors and relevant entities.

**Breakdown of Project Activities for Schedules**

1. Breakdown of activities for schedule is consistent with project estimate.
2. Breakdown of activities is in compliance with contract requirements.
3. Breakdown of activities is consistent with varying difficulty of work activities; breakdown is in enough detail.
4. Breakdown of activities is not too detailed to enable job site monitoring and updating of work activity.
5. Breakdown of activities is consistent with project control system.

## Relationship Between Project Activities for Overall Schedule

1. The activity relationships correctly illustrate technical relationships; e.g., walls are constructed subsequent to the footing put in place.
2. The activity relationships correctly consider resources available to the project; e.g., two activities going on at the same time require a sum of 12 workers and only 10 are available to the project.
3. The activity relationships correctly consider preference or economic considerations of the building process; e.g., the schedule shows two activities going on at the same time when in fact the contractor intended to and performed the activities one after the other.
4. The activities reflect the proper type of logic between them; finish to start, finish to finish, or start to finish.
5. The logic between any two activities recognize the required lag time; e.g., time for concrete curing.

## Correctness of Activity Durations

1. The calculation of activity durations recognize the quantity of work that must be done for an activity.
2. The activity durations do properly reflect the crew size (labor and equipment) that will be utilized to perform the work.
3. The activity durations do properly reflect the productivity of the crew size that will be performing the work.
4. The activity durations reflect historical data regarding productivity for the work activities.
5. An activity duration does reflect the impact of supervision or resource requirements of activities that will be performed at the same time.
6. Activity durations do properly reflect the time or project constraints.
    a. Temperature (cold, hot, precipitation).
    b. Project logistics.
    c. Various project milestone dates.
    d. Uniqueness of work or specification tolerances.

## Calculations and Use of Schedule

1. Schedule calculations to include CPM calculations are performed accurately.
2. There are internal controls to check computer or manual schedule calculations.
3. The specific schedule is prepared and submitted and utilized; e.g., the earliest start time schedule is submitted but the intent is to perform the project schedule on a late start time basis.
4. The project schedule is communicated or distributed to all relevant project entities; e.g., to project personnel.

5. The schedule is broken down into achievable segments to permit an effective means of scheduling and assigning resources to include workers.

# Appendix C

# Implementing a Productivity Program

## Productivity Improvement Step

**Firm Evaluated/Name** _____ **Date** _____

1. A reasonable labor productivity estimate is prepared and used as a benchmark goal by job site supervisors.
2. The contractor uses a job control system to monitor and detect labor productivity problems/overruns that have the following characteristics:
   a. timely (weekly/monthly reports)
   b. compares estimated productivity (work put in place and labor hours expended) to,
   c. actual productivity (work put in place and labor hours expended)
   d. with the objective of "detecting" any and all productivity overruns and identifying the cause,
   e. specific timely steps are taken to correct problem(s).
3. The contractor employees qualified supervisors (project managers, superintendents, foremen).
4. Continuity of supervisors are maintained throughout the project.
5. An accurate project master schedule is prepared, updated, and utilized by job site supervisors as a means of effectively scheduling work, resources, and making corrective actions when unexpected events take place.
6. A one to three week formalized "look ahead" schedule is done in conjunction with the updated master schedule to enable effective scheduling of crews and workers.
7. A formalized daily planning and scheduling form or system is utilized to schedule crews, workers, and to procure needed equipment, tools,

and materials.

8. Manpower is scheduled to provide a smooth hiring and laying off of workers as a function of project time.
9. Competent skilled workers are hired and evaluated throughout the project.
10. Competent foremen are utilized that lead workers, provide needed direction to workers, and monitor workers.
11. A sufficient number of job site supervisors are maintained on the project relative to the changing number of craftsmen as a function of project time.
12. The contractor provides needed equipment and tools for workers to perform work efficiently.
13. The contractor continually analyzes job logistics to enable an effective storage of materials, tools, and equipment.
14. The contractor takes steps to keep the morale of workers positive to include:
    a. keeping workers informed of work progress.
    b. seeking input from workers on productivity improvement and means of correcting problems.
    c. implementing a program of pride of work.
15. The contractor takes steps to minimize the amount of redo work that workers have to perform.
16. The contractor takes steps to minimize the amount of time workers "wait on assignment" owing to the fact that the workers were done with a task and did not know what to do next.
17. Job site personnel attend daily/weekly planning meetings at which schedules, and problem areas are addressed.
18. A quality program is implemented to reduce/eliminate the need for punch list work.
19. The contractor effectively coordinates schedules and the work plan with other project contractors.
20. The contractor's project organizational structure is designed and executed efficiently (lines of authority, decision making responsibility, etc.).

# Appendix D

# Job Cost Control Procedures

## Job Cost Control Issue

**Firm Evaluated/Name** _____ **Date** _____

### Integration with the Estimate, Schedule, and Accounting System

1. The cost codes or work breakdowns are consistent with the work estimate used to bid the project.
2. The cost codes or work breakdowns are consistent with the project schedule breakdown.
3. The job cost system does tie to the general ledger accounting system.
4. There is no missing data or job cost reports owing to a change in the job cost system or accounting system.
5. There is evidence that the amount of work billed to the project owner does correlate to progress reported in the job cost reports.

### Format of Report

1. The cost codes or work breakdowns are such that quantities of work put in place can be effectively measured at the job site for each reporting period.
2. The cost codes for the work codes are defined such that there is a one to one correlation between the quantity of work performed and the labor hours expended to accomplish the work.
3. The structure of the job cost report is kept constant throughout the project.

4. The cost report is such that quantities of work put in place or labor hours expended that do not fit predefined codes, are identified to a new code or are segmented.
5. The job cost system is not such that labor hours are only compared against the budgeted labor hours and a problem is detected when the actual hours exceed budgeted hours.
6. The job cost work codes are such that they do reflect the differences in the difficulty or work effort for various types of work required.
7. The job cost system does project variances or overruns based on the progress to date.

**Accuracy and use of data input**

1. Job site personnel to include foremen are properly instructed how to fill out time cards and quantity reports.
2. Time cards and quantity reports are filled out consistently between the workers; e.g., each foreman fills out time cards to the same degree of detail.
3. Management and administrative personnel effectively monitor and correct the inputting of job cost input data from the field.
4. Time cards and daily reports are legible such that they can be input to the job cost system.
5. The job site data is input/transferred into the job cost system in a timely and accurate process.
6. The job cost report is reviewed by management and job site supervisors in a timely manner to enable the identification of problem areas.
7. The job cost report is reviewed by management and job site supervisors in a timely manner to enable the identification of opportunities for improvement.
8. Procedures are in place to have job site supervisors react and account for productivity problems that are indicated on the job cost report.
9. Procedures are in place to back up the job cost reports with copies kept at the main office or at a secure site.

**Job Cost Report and Change Order Work**

1. Change order work is segmented from base bid work in the job cost work codes.
2. Change order work is handled in a manner that the actual quantities and labor hours expended are updated in the cost reports.
3. Time cards and quantity reporting forms are such that job site personnel can adequately segment change order work (quantities and time reporting).

4. Procedures are in place to differentiate between the reporting and segmenting of quantities and labor hours for approved change order work
and nonapproved or disputed change order work.

# Appendix E

# Claims Prevention and Preparation

## Claim Prevention/Preparation Issue

**Firm Evaluated/Name** _____ **Date** _____

**Preparation of Estimate**

1. In preparing a project estimate, the contractor establishes an hourly rate for use of his owner-purchased equipment and budgets a cost to the job for each piece of equipment to be used at the job even if the equipment is full depreciated.
2. To the degree possible, the contractor utilizes his own historical labor productivity data from past projects to prepare a new project estimate and includes a "trail" in the job-site record file giving an explanation of the interface between the project estimate and his historical data.
3. Job site overhead items, such as job-site trailers, small tools, and required supervision, are budgeted as individual items in the project estimate to facilitate a comparison of actual costs versus budgeted costs for these items.
4. A checklist is used that reminds the contractor to search for every work item as part of the quantity takeoff process to ensure that various types of items are not missed in the takeoff.
5. A checklist is used that reminds the contractor to include every type of job overhead in the estimate, including a specific list of possible types of insurances to be included.
6. As part of the estimating of the direct labor costs for a project, the contractor makes an analysis of the assumed crew sizes and number of craftsmen to be assigned to specific work items and the project.

7. As part of the estimating of a new project, the contractor attempts to analyze the amount of company overhead costs that will be required to support the project being estimated.
8. Procedures are in place that result in double checking of mathematical calculations in the estimate.
9. As part of the process of determining the profit to be added to a bid, the contractor takes into consideration the expected weather conditions for the period in which the project is to be constructed.
10. All assumed project start dates and milestone dates established by the project owner or his designer should be written down by the contractor and considered in his profit determination for a project such that the contractor can quantify a damage claim should he be delayed beyond one of these dates.
11. When the contractor is required to submit a project schedule, any project float or slack on the schedule is put into a profit consideration such that the contractor can quantify damages if an event outside his control should result in his losing some of this assumed float.

## Preparation of Project Plans and Schedules

1. If the project owner requires the contractor to submit a formalized schedule, the contractor submits one that is acceptable to the owner, but also sets out the contractor's conservative estimate; e.g., a latest start time schedule.
2. When required to submit a project schedule for the project owner or designer, the contractor does not submit a schedule that has a completion date earlier than the required completion date set out by the owner or designer.
3. If the project owner requires the contractor to submit a project schedule and revise and submit it to the project owner or designer periodically as the project progresses, the contractor adheres to the requirement without exception even if the project owner or designer is not strict in enforcing the requirement.
4. When required to submit a project schedule to the project owner, the contractor assigns durations to the individual activities defined on the schedule that "tie" to the quantity takeoff amount of work to be performed, the planned labor assigned to the activities, and the expected productivity.
5. When a project schedule is required by the project owner, the contractor takes steps to review the schedule to determine that the technical relationships of the activities illustrated are possible.
6. The contractor should not submit a project schedule that commits the contractor to defined project progress unless the project owner or designer requires submittal of such a plan.

7. When required to submit a project schedule to the project owner, the contractor takes steps to ensure that he has recognized resource availability in his project activity logic.

8. When required to submit a project schedule, the contractor takes steps to include "preference" logic in the project schedule. For example, even though it may be possible to form some concrete slabs while placing other slabs, the contractor may prefer to show these as one activity following the other, as he may believe that it is more economical to perform them this way.

9. When required to submit a project schedule to the project owner or designer, the contractor should take a conservative view in setting out the amount of float/slack he illustrates following project activities on the schedule.

## Record Keeping for a Project

1. The contractor documents, via a written log, all pertinent instructions and work assumptions made orally by the project owner or designer at preconstruction and job site meetings.

2. Every day at the job site, job-site personnel are required to document the quantity of work put in place for each work item/activity.

3. Each and every day at the job site, job site personnel are required to document the degree of effort expended per work item to include the number and type of craftsmen working on each work item.

4. Every day at the job site, job site personnel are required to document the factors that characterize the work that day, including weather, owner or designer instructions, subcontractor interference, owner purchased material, etc.

5. When events occur that may cause a scope of work or quality dispute regarding the work performed, the contractor takes photographs or a film of the work in question.

6. Every day at the job site, job site personnel document all questions and inquiries made to the project owner or designer and also the responses.

7. The contractor records daily the nonperformance of other contractors, which may lead to the contractor's having to change his performance or dates.

8. In the daily records maintained at the job site, the contractor's personnel should document and explain any instances in which the contractor believes the design firm's representative is overinspecting the contractor's work.

9. In the daily records maintained at the job site, the contractor's personnel should document any instances in which the contractor believes the design firm is attempting to "direct" the work methods of the contractor.

10. In the daily records, the contractor records any instances where he believes the project owner or designer is accelerating the contractor's work progress (e.g., requiring overtime or increased crew sizes).

## Change Order Work

1. The contractor does not proceed with change order work until the project owner or designer has agreed to a scope of work and an amount to be paid to the contractor for performing the work.
2. When doing change order work, the contractor creates a new cost code in his accounting to capture the cost of doing the change order work.
3. All costs and documents regarding a specific change order are kept in an orderly fashion in a separate job cost file in the main office.
4. When required to perform change order work, the contractor immediately calculates the possible impact on the project schedule and requests additional time for the work to include a needed project date extension.
5. To the degree possible, job site overhead costs related to performing change order work should be traced to the change order work.
6. To the degree possible, even company overhead costs related to performing change order work should be traced to the change order work.
7. When requested to perform change order work, the contractor should attempt to quantify any lost productivity impact that the change order work might have on the other work activities and immediately seek compensation for this impact as part of the change order amount.

## Claims Administration

1. All management personnel to be employed at the job site should have been trained in the claim's prevention procedures the firm practices.
2. When confronted with a potential claim event, the contractor should immediately inform the project owner or designer of the situation and make a sincere effort to resolve the dispute or claim in good faith with the project owner or designer as soon as possible.
3. Given the occurrence of a claim event or damage, the contractor should take steps to minimize the damage incurred and document this as evidence to enable the contractor to show "good faith".

# Index

overlooked accounting entries, 121
overtime, 184, 267, 274
owned equipment, 191, 238, 288
owned-equipment use, 271
owned-equipment-cost claim, 275
owner, 35, 173
owner–agent relationship, 173
owner-independent contractor
    relationship, 173
ownership, 92, 192
ownership rates, 193
owning equipment, 246

paperwork, 264
par market value, 89
par-value, 84
parol evidence, 157
participation loan, 66
partnership, 14, 16–18, 20, 82, 85, 158
partnership agreement, 83
partnership income, 17
partnerships, 72
past performance data, 192
past project productivity, 188
pavers, 88
pay request procedures, 218
payables, 100
payee, 169, 171, 172
payment in full, 169
payment process, 141
payment retention, 135
payroll, 95
payroll checkoffs, 184
payroll function, 27
penalty charge, 60
penalty charges, 172
penalty costs, 163
Pennsylvania System, 166
percentage complete, 105
percentage of completion, 102
percentage of effort, 251, 252, 258
percentage of work, 251, 252, 258
percentage ownership, 30
percentage-of-completion, 138
percentage-of-completion method, 93,
    101, 103, 107–109, 113, 139
performance management, 258, 259
performing CPM calculations, 214

period revenue, 90
periodic depreciation, 90
perpetual inventory system, 247
perpetual system, 115
personal contact, 50
personal contacts, 51
personal property lien, 166
personnel management, 35, 228
personnel management program, 227
petty cash, 87
petty cash account, 87
petty cash disbursements, 88
petty cash fund, 87
petty cash funds, 87
physical commodities, 35
physical inventory, 115
placing, 53, 54, 55
placing aspect, 53
planned, 257
planned duration, 257
planned non-productive time, 208
planned work, 274
planning, 34, 36, 54, 55, 208, 217, 258
planning and procurement functions,
    27
planning and scheduling, 5, 8, 10, 234,
    245, 294
planning for company operations, 36
planning/scheduling, 206
plug number, 122
police officer, 150
pooled rate, 242
poor communications, 229
positive worker attitudes, 228
potential profit, 178
power to tax, 150
pre-planning time, 231
precedence diagram, 213, 214
preference logic, 212
preferred stock, 73
premium-on-capital stock, 84
prepaid accounts, 118
prepaid expenses, 114
prepaid income, 114
prepaid lease expense, 93
preparation of financial statements,
    113
prepared estimate of costs, 180